D1571338

THIN FILM PHYSICS

THIN FILM PHYSICS

O. S. HEAVENS

D.Sc., F.Inst.P., F.I.E.E.

Professor of Physics
University of York
Heslington, York

METHUEN & CO LTD

11 NEW FETTER LANE LONDON EC4

First published 1970

Filmset by Keyspools Ltd
Golborne, Lancs
Printed in Great Britain
by T. & A. Constable Ltd
Edinburgh
SBN 416 07650 5

Distributed in the United States of America
by Barnes and Noble Inc.

Contents

Preface

Although some nineteenth-century activity is recorded in connection with thin film studies, it was not until the middle of the twentieth century that a really substantial effort began to be devoted to this field. There are two main reasons for this. On the one hand, the experimental techniques required in order to carry out meaningful studies had not been developed. On the other hand it was found that many of the rapidly increasing demands of technology could be met in a thorough and elegant fashion by the exploitation of thin film properties.

This monograph provides a fairly extensive view of the present state of progress in a number of important areas of thin film physics. A complete book could be written on any of the chapters contained herein – in fact several such books already exist. Thus the present work is not intended to provide in-depth studies of any particular area, but rather to provide, for the general reader or for the newcomer to the field, a general background from which interests in individual directions may develop. As will be seen, the state of progress in different fields varies somewhat. In some cases, highly sophisticated models have been evolved to account for detailed and subtle behaviour of films: in others, only rather general comments are possible.

Within the last few years, the improvements in experimental techniques have begun to exercise a decisive influence on the progress of thin film physics. For the first time, experiments have become possible under very highly controlled conditions. Diagnostic techniques have been developed which enable far more direct information to be obtained than hitherto. In some of the sections of this book, the impact of these new methods is beginning to appear.

It is a pleasure to acknowledge the assistance of Drs Prutton, Chambers and Gallon, for providing the diffraction patterns for Plates I, III, V, VII and VIII, and to thank the Optical Society of America for permission to reproduce fig. 6.7.

1

Introduction

At the time of writing, a very considerable part of our industry is involved with thin films. Much of this is old-established, as in the use of electroplated films for protection or decoration. Much of it is of recent origin and involves the use of techniques which have been developed over the last few years. In one way and another most physical properties of films – optical, chemical, magnetic, electrical, etc. – are of importance in an ever widening sphere of industrial, scientific and technical applications. At the same time studies of the fundamentals of film formation and of the basic reasons for differences in behaviour between films and bulk materials are being pursued with increasing vigour. With the development of sophisticated methods of production and examination, our understanding of many of the apparent vagaries of film behaviour is steadily improving. Many of the discrepancies in our early experiments on films are now known to have arisen from factors whose influence was unknown or unsuspected. Advances in many of the necessary accompanying technologies have enabled work on films to be carried out under conditions admitting a high degree of control.

The first results of such recent developments have shown an increase in the reproducibility of results obtained on *some* thin film systems. Let us not give the impression that no problems remain to be solved in this respect. As will be seen in the discussions of the succeeding chapters, there remain several fields in which, in spite of the most careful control of conditions of preparation, wide variations in film properties result. In some cases, this characteristic remains an effective barrier to the commercial use of films in developments where it is known that great advantages could accrue.

Our problems in studying thin film systems begin before we produce the film. The structure and properties of many films are known to depend considerably on the state of the surface on which they are deposited. It is to be expected that this will be so for films of average thickness corresponding to only a few atomic diameters, and this is indeed found to be so. Sometimes, however, the influence of the substrate makes itself felt in films of considerably greater thicknesses than this. The way in which the crystallographic form of films depends on substrate surface conditions is discussed in Chapter 4.

Our initial problem, then, in dealing with film properties is one of surface physics. We need to know exactly what kind of surface is being used for the deposition of the film – whether it is crystallographically oriented or not; what density of dislocations or imperfections exists; whether any absorbed gas is present and if so, what gas and how it is bonded to the surface; if the surface is gas-free, then what is the spatial distribution of surface atoms of the substrate and what is that of the electrons in the neighbourhood of the surface atoms; what are the activation energies for desorption and surface migration of the deposit atoms on the surface. These are typical of the questions to which answers may be needed before an adequate description of film properties can be made. We are, in most cases, not able to furnish even a fraction of the answers. The reason for this is in part, that the necessary experimental techniques for finding many of the answers have only recently come into our hands. The results of the application of some of the powerful techniques recently developed have sometimes been surprising. Thus we had happily assumed that the crystallographic arrangements of atoms on the surface of a crystal would be not too violently different from that of a parallel plane inside the crystal. We would expect that the interplanar spacing normal to the surface may differ from that inside the crystal. We may even imagine that the surface atoms may be displaced along the normal to the surface by different amounts, depending on their crystallographic position. We would *not* expect to find the symmetry of the distribution of the surface atoms to differ radically from that of the atoms inside. This, however, is precisely the kind of thing which has been observed when the recently developed technique of slow-electron diffraction coupled with post-diffraction acceleration has been applied. Shattering observations of this kind make it clear that, even with the clean surfaces which can be produced and maintained in the present ultra-high vacuum system, we are still some way from being able to discuss with confidence the detailed mechanics of condensation phenomena.

In studies of the early stages of film formation it is often not clear whether nucleation or growth effects constitute the dominant mechanisms determining the film structure. Macroscopically it is clear, from electron microscopy, that the film in the sense of a uniform, continuous distribution of atoms, is a rarity, if indeed it ever occurs. However, such observations are based either on films grown in what we now dismiss as 'dirty' vacua – say in the region of 10^{-5} torr – or on films which have been exposed to the air. Our only present method of examining the surface distribution of absorbed atoms under clean conditions is by the indirect method of slow-electron diffraction. Until some intrepid instrument designer produces a high-resolution electron microscope operating at a pressure of 10^{-10} torr, we must remain in semi-darkness on this point.

In this book, we shall devote most of our attention to films whose thickness is large compared with the interatomic spacing. In many cases (although not all) such films exhibit behaviour which is at most only mildly influenced by the underlying substrate, at least in the sense of being more or less independent of the effects of the binding forces to the substrate.

We shall first review the various methods by which thin films are produced, discussing the respective merits and demerits of the methods described. The range and scope of methods used to examine film properties and structures will be dealt with and will be followed by an account of the types of structure obtained under various conditions of preparation. The ensuing chapters will then deal specifically with the physical properties of films which have been the subject of study. These chapters will also include accounts of the more important applications of film systems.

2

Methods of Preparation of Films

The methods commonly used to prepare thin films of solids may be classified under the following headings:

(*i*) Electrolytic deposition, which includes cathodic and anodic deposition;
(*ii*) Sputtering, including the more recently developed method of reactive sputtering;
(*iii*) Deposition in vacuum from a heated source (thermal evaporation);
(*iv*) Deposition from vapour reactants;
(*v*) Diffusion.

In some cases, method (*iii*) may be used reactively, by introducing a gas which reacts with the evaporated material. The main characteristics of these methods are given below.

2.1 *Electrolytic deposition – cathodic films*

The method of making metal coatings by electrolytic deposition on the cathode of a cell is probably one of the oldest known methods for making films artificially. Superficially, the method is simple. Ions in the solution are impelled to the electrodes by the applied electric field. At the electrodes, the charges are neutralized and straight deposition may occur. Alternatively reactions may occur between the resultant atoms or radicals and the electrode or bath materials. The laws governing the overall process are well-known but factors governing the precise form and microstructure of films formed in this way are less well understood. As the method has been developed industrially, so valuable empirical techniques have been evolved which enable considerable control to be exercised, e.g. on the brightness, durability, etc. of the films. In many cases the detailed mechanisms of operation of many of the additives used are not yet well understood.

A simplified view of the electrolytic deposition process, inadequate in detail but giving an indication of the main factors influencing the process, may be seen from fig. 2.1. A double layer forms at the cathode of the system, formed by the electrons in the metal and the ions adsorbed on the surface. In this region, the dominant forces are of short range. Adjacent to this region is a diffusion layer, lying between the cathode and the main body of the electrolyte. In the latter, the electrolyte concentration is substantially constant and the transport of ions is effected by the applied electric field. In the cathode diffusion layer, the combined effects of the concentration gradient and field apply so that the motion of the ions is governed both by diffusion and by electrical forces.

From fig. 2.1 it may be deduced that such factors as temperature and state of motion of the electrolyte will, through their influence on diffusion of the ions, affect the deposition conditions. The geometry of the deposition system will play a part since this will determine the local current density. The presence of the double layer on the cathode surface means that there will be a marked selective effect on the adsorption of solution constituents. Thus a significant degree of control is available through the use of additives which exercise buffering effects or inhibit bubble formation, and through which the stress, brightness, texture and other properties of the electrodeposited surface may be determined.

2.2 *Electrolytic deposition – anodic films*

When certain metals are made the anodes of electrolytic cells, oxide film formation occurs, leading frequently to extremely hard, compact well-adhering coatings. Anodised aluminium articles have long been generally available – excellent examples of the way in which the normally highly reactive aluminium is completely protected by the anode film. The electrolytic capacitor forms another everyday example of the use of such films. In the field of electron microscopy, use has in the past been made of anodic aluminium

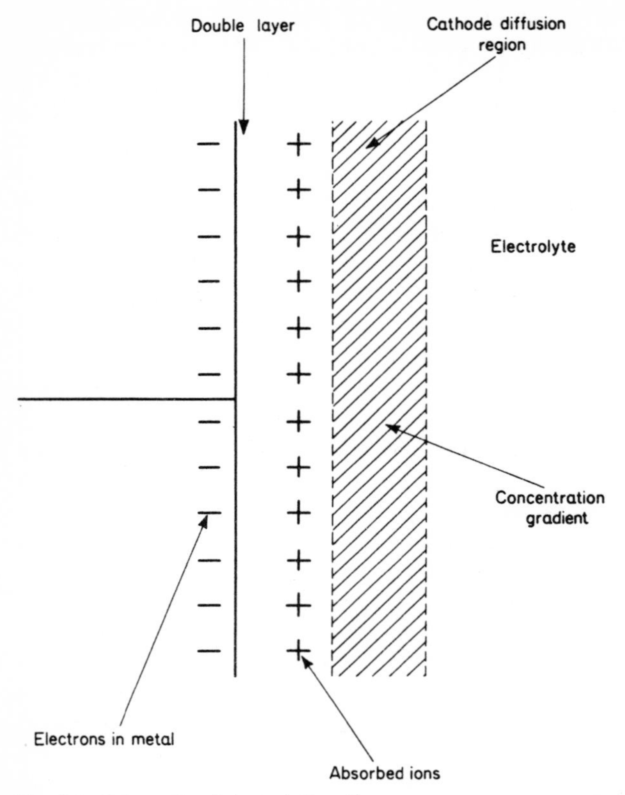

FIG. 2.1. Schematic of electrolytic cell.

oxide as a substrate for transmission electron microscopy. After removal from the metal, the oxide is self-supporting in thicknesses down to less than 10 nm and is amorphous, so that no appreciable detail is seen in the micrograph or diffraction pattern.

The metals on which anodic films may be formed are:

Aluminium	Hafnium	Tin
Antimony	Magnesium	Titanium
Beryllium	Niobium	Tungsten
Bismuth	Silicon	Uranium
Germanium	Tantalum	Zirconium

A typical current/voltage relation for the anodic process is shown in fig. 2.2. At any applied voltage, a combination of electron (leakage) current and ionic current flows although in the region OA, the ion current is negligible compared with the leakage current. Once an appreciable ion current flows, then the behaviour becomes time-dependent, since oxide film is being formed so that the barriers to current flow are changing. It is convenient to consider separately (*a*) growth at constant current and (*b*) growth at constant applied voltage.

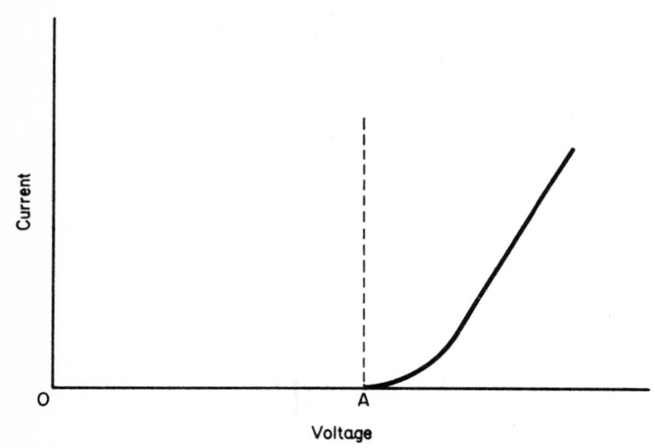

FIG. 2.2. Current vs voltage for anodic process.

It is assumed that the structure and composition of the film is constant during formation and that a constant differential field strength $\mathrm{d}V/\mathrm{d}l$ is required (V = applied voltage, l = film thickness) to maintain a constant ionic current. Under conditions where the electronic leakage current is negligible, the rate of increase of film mass is proportional to the current so that, under constant current conditions, $\mathrm{d}l/\mathrm{d}t$ is constant. The voltage/time relation is thus linear since

$$\frac{\mathrm{d}V}{\mathrm{d}t} = \frac{\mathrm{d}V}{\mathrm{d}l}\cdot\frac{\mathrm{d}l}{\mathrm{d}t}.$$

In practice a slight dependence of $\mathrm{d}V/\mathrm{d}l$ on current is observed. For tantalum at 18° C, the value of $\mathrm{d}V/\mathrm{d}t$ for a current density of 10 mA.cm^{-2} is about 3.6 volts.sec^{-1}.

In applications of the method of anodic film growth, a constant potential is generally applied. Since the growth of the film results in a diminished driving field, the current decreases with time. In practice, the rate of increase of current and hence of thickness falls to a very low value. In fact, the growth rate always remains finite, so there is no question of the film's growing to a limiting thickness – the film will continue to grow indefinitcly. However it is found that with, e.g. aluminium, application of a fixed voltage V to a cell rapidly produces a film of thickness $\sim$ 1·3 V nm and that the rate of increase beyond this value is very low indeed – probably such as to show no measurable increase over periods of the order of an hour or two. Thus the concept of an approximate growth/voltage constant is useful in this sense.

The detailed mechanism of film growth is one of some complexity. The basic requirement is that of a diffusion mechanism for ions through the film and it is generally assumed that it is the metal ions which travel through the oxide. It is clear that the metal/oxide interface must during growth be in a somewhat dynamic equilibrium. Little is usually said of the way in which the film bonds to the metal under these conditions, for the very good reason that little is known. For discussion of the kinetics of the growth process it is assumed that metal ions pass through the oxide by interstitial diffusion. If it is assumed that the variation of the potential energy of

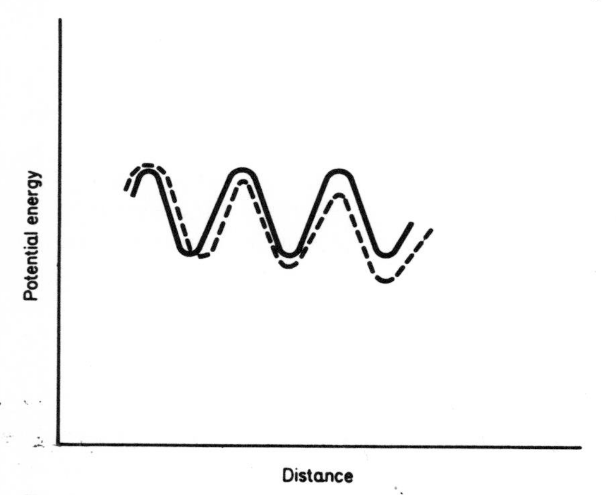

FIG. 2.3.

the ion through the lattice in the field direction is as shown in fig. 2.3, then the effect of an applied field E on the motion of an ion with a charge Ze is to reduce the zero-field barrier W by an amount $ZeaE$. The probability that an ion will surmount the barrier of height $W-ZeaE$ is given by

$$\nu \exp\left\{-\frac{(W-ZeaE)}{kT}\right\}$$

so that the current density in the forward direction will be

$$i_{+} = 2an\nu \exp\left\{-\frac{(W-ZeaE)}{kT}\right\} \tag{2.1}$$

where n is the ion density, which is in general a function of l, and ν is the oscillation frequency of the ion in the lattice. For the reverse direction, where a barrier of height $W+ZeaE$ is involved, the current is

$$i_{-} = 2a\left\{n+2a\frac{\partial n}{\partial l_{a}}\right\}\nu \exp\left\{-\frac{(W+ZeaE)}{kT}\right\}. \tag{2.2}$$

Two extreme cases may be recognized. If $ZeaE \gg kT$ ('high-field' case), then i_{-} is negligible compared with i_{+} so that equation 2.1 holds. If $ZeaE \ll kT$, then the net current $i = i_{+}-i_{-}$ is given, to within this ('low-field') approximation as

$$i = 4a^{2}e^{-W/kT}\left\{\frac{n\nu ZE}{kT}-\frac{\partial n}{\partial l}\right\}. \tag{2.3}$$

A refinement of the above simple picture is required to allow for the fact that the potential barrier to removal of the metal ion from the metal will be different from that for diffusion between interstitial sites. Thus the potential curve of fig. 2.4 is more appropriate. If we assume that v = atomic volume of the metal and Ω that of the oxide (i.e. a volume v of metal produces a volume Ω of oxide), then for unidirectional, high-field conditions, the rate of increase of film thickness for a current density i will be given by

$$\frac{dl}{dt} = \left(\frac{iv}{Ze}\right)\Omega \exp\left\{-\frac{(W-ZebE)}{kT}\right\} \tag{2.4}$$

where $E = V/l$, the field in the oxide and l the oxide thickness. If we adopt Cabrera and Mott's (1949) criterion that the limiting growth rate is that corresponding to the addition of one atomic

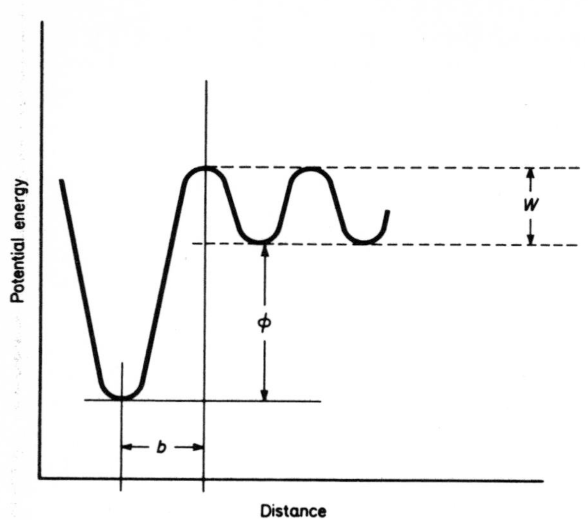

FIG. 2.4.

layer per day, so that $dl/dt \sim 10^{-13}$ cm.sec^{-1} we can determine the limiting thickness L at which this occurs – viz:

$$L = (ZbV/kT)\left[\frac{W}{kT} - \ln(10^{-13}/A)\right]^{-1} \qquad 2.5$$

where $A \equiv Ze/iv$. Inserting plausible values for the parameters involved in this equation, a limiting thickness for the thermal oxidation of aluminium is about 4 nm, in reasonable agreement with the experimentally observed thickness.

If the Mott-Cabrera theory is applied directly to the case of anodic oxidation, it leads to predictions which accord poorly with the experimental observations. A logarithmic law is observed experimentally (cp. equation 2.5) but numerical values of the constants are seriously in error. Moreover the experimental results suggest that the term $ZebE/kT$ is independent of temperature, implying that b varies linearly with absolute temperature. On physical grounds, this appears unlikely.

An omission of the potential barrier theory outlined above is that it does not take account of tunnelling at the barriers. In certain circumstances, tunnelling may be significantly more probable than thermal excitation over the barrier. Thus for the case of zirconium, for which the barrier height is deduced as 0·375 eV with a width $2a = 0{\cdot}72$ nm, the probability of thermal excitation for a field of $F = 5\times10^8$ volts.metre^{-1} is $\sim 4\times10^{-4}$. The tunnelling probability $P(F)$ for this field is given by

$$P(F) \sim \exp\left\{-\frac{2}{h}\int_{l_1}^{l_2}[2m(V-F)]^{\frac{1}{2}}\mathrm{d}l\right\} \qquad 2.6$$

which, for the case considered, is $\sim$0·05. Although a crude model, this serves to illustrate the relative importance of the two mechanisms.

Two further effects which indicate that the simple theory given above needs extension are:

(*i*) At low fields, experimental results tend to follow a linear $\log i$ vs $E^{\frac{1}{2}}$ law rather than the linear one predicted; (*ii*) at high fields ($\sim 6\times10^8$ volts.metre^{-1}) a discontinuity is observed in the plot of activation energy vs field. The first-mentioned difficulty may be overcome by the assumption that a low (10^6 cm^{-3}) density of electron traps exists in the oxide. The second point is consistent with a change of form of the potential/distance curves with increasing field strength, whereby at low or moderate fields, the highest part of a barrier may be at the second or third maximum from the metal surface whereas at high fields, the first maximum is the highest. In this case one would have a discontinuity in the activation energy.

At this time, there is no single complete theory to account for the vast number of results in this field. The degree of success with theories such as those discussed above is remarkable when one remembers that most of the oxide films under discussion are glass-like rather than crystalline. There will therefore be a whole range of site types, jump distances, activation energies, all of which introduce difficulties in formulating a complete theory. Moreover, many of the experimental results may sometimes be the spurious results of trace impurities whose influence is as yet not understood.

2.3 *Sputtering*

The method of film deposition, whereby a d.c. glow discharge is established between a plate of source material and the substrate, was used as early as 1842. Many of the interferometer plates used by spectroscopists of the last century were coated by sputtering.

Although the experimental arrangements required are simple, the process is a somewhat complex one. Basically removal of the cathode material results from the bombardment of the cathode by energetic positive ions of the discharge. Although there are many features of the sputtering process which remain to be understood, it now seems well established that the removal of cathode material is associated with momentum transfer from the bombarding ions. In simple sputtering, no chemical reactions are involved. The mean velocity of the sputtered atoms arriving at the substrate may be quite high – far higher than that of atoms from thermal evaporation sources, described in the next section. The implications of the high energy of the condensing atoms on the structure of the films formed may be wide. It is known, for instance, that sputtered particles may penetrate to depths of the order of a few nanometres in solid substrates. Thus it is to be expected that damage to the substrate surface will occur, and it is possible that the sputtered film may be anchored to the substrate by 'roots' of deposit atoms, penetrating the damaged surface.

Sputtering is generally carried out at pressures of the order of a tenth of a torr of carrier gas. Measurements of various physical properties of films produced by sputtering revealed, for films produced in early days, wild discrepancies. It seems likely that these can be to some extent ascribed to impurities in the carrier gases used since in more recent work, for which the chamber is first pumped out to a high vacuum followed by sputtering in a gas of high purity, more reproducible results are obtained.

An indication of the uncertainties which can arise from impurities in the sputtering gas may be seen in the case of films of nickel. The normal crystallographic form of nickel obtained in films produced by thermal evaporation is cubic. In many early experiments on sputtered nickel, using argon as the carrier gas, a hexagonal form of nickel was reported. Significantly, this was never obtained when spectroscopically pure gas was used. Two further points are relevant: (*i*) the reported nickel structure possessed lattice constants suspiciously similar to those of nickel nitride and (*ii*) nitrogen is a common impurity in the commercial grade argon used in these experiments. In fact a hexagonal form of nickel is known to exist and may be produced by electrolysis under suitable conditions. The lattice constant is, however, quite different from that obtained for the sputtered 'nickel' and is a more plausible one when the interatomic distances are considered.

2.4 *Thermal evaporation* in vacuo

Interest has grown over the past 30 years or so in this method of preparing films. The pressure of gas in a chamber is reduced to as low a value as possible and the source material is heated to a sufficiently high temperature to ensure that evaporation (or sublimation) will occur at a suitable rate. For the purpose of obtaining a 'parallel-sided' film on a plane substrate, it is sufficient to ensure that the pressure of residual gas in the chamber is low enough for the mean free path of the atoms of evaporated material to be larger than the source-substrate distance. In fact, films produced under conditions where this is only marginally satisfied may well show a marked influence of the residual gas in the chamber. Some comments on the dependence of film properties on the residual gas will be found in the chapters dealing with specific physical properties of films. Suffice it to say that highly misleading results may be obtained.

A wide variety of evaporation sources has been developed for the thermal evaporation method. For a comprehensive account of these, the reader is referred to Holland's (1956) excellent book on the subject. A representative selection of source types is given below.

Shaped boats of refractory metal, such as tungsten, tantalum or molybdenum serve for a wide variety of materials, including many non-metals and those metals which do not dissolve the boat material. If solution of the boat material is a problem a carbon boat may be useful or, for a low melting point material (e.g. Se), a porcelain boat set in a tungsten outer boat. Metal boats of this type are conveniently heated by passing a current through them. When chamber pressures of the order 10^{-5} torr are used, there may be parts-per-million contamination of films by the boat material, even though the solubility of boat metal by the charge is practically zero. This may well arise from transfer of boat material by reaction with the residual gas. Thus transfer of tungsten may occur through the historic 'water-cycle' in which tungsten reacts with water vapour to produce tungsten oxide and atomic hydrogen. The tungsten oxide evaporates and is subsequently reduced to tungsten by the hydrogen, thus releasing more water molecules to continue the process.

Crucibles of refractory oxides or carbon are sometimes used, either heated by electron bombardment (fig. 2.5) or by induction

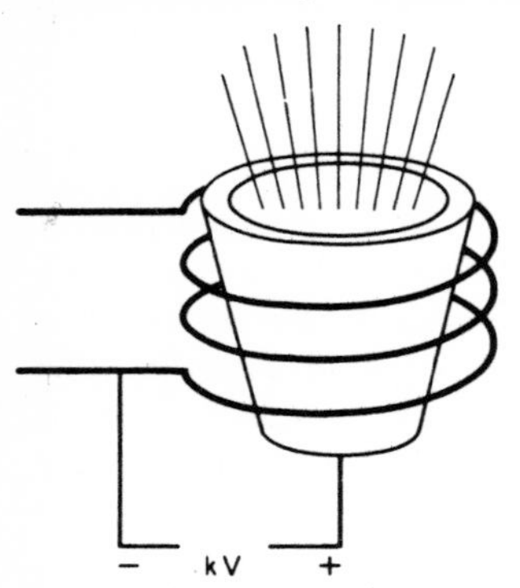

FIG. 2.5. Crucible heating by electron bombardment.

heating of the (metallic) charge. Some difficulty was experienced in devising an evaporation system suitable for silicon, which tends to attack many of the usual crucible materials. This problem was solved by the arrangement shown diagrammatically in fig. 2.6.

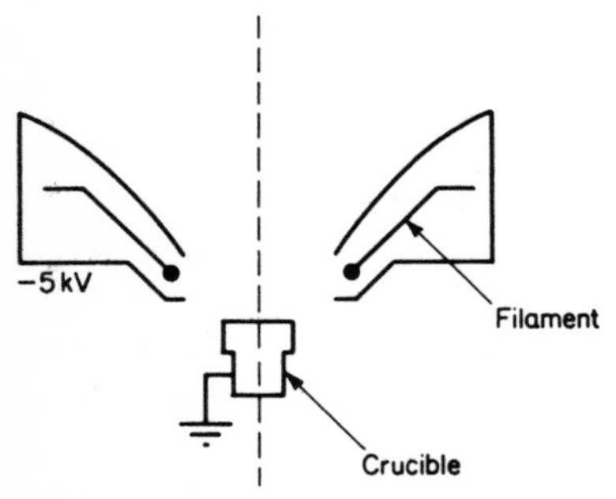

FIG. 2.6. A typical ring gun assembly.

Simple electrostatic focussing of the electrons emitted from the tungsten filament produces an intense electron beam at the silicon charge. This rests on a water-cooled support. It should be noted that the target surface cannot see the tungsten filament so that contamination from this source is avoided. Alternatively, the end of a rod may be heated by electron bombardment (fig. 2.7) to form a pendant drop. In this case, contamination by crucible material is

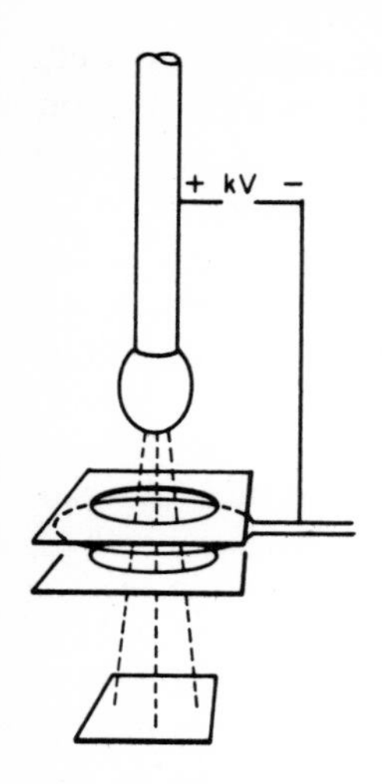

FIG. 2.7. Electron bombardment of pendant drop.

completely eliminated since there is no crucible. (This method yields an unexpected and irrelevant bonus – if the pendant drop is photographed, the surface tension of the metal may be found from measurements of the shape!)

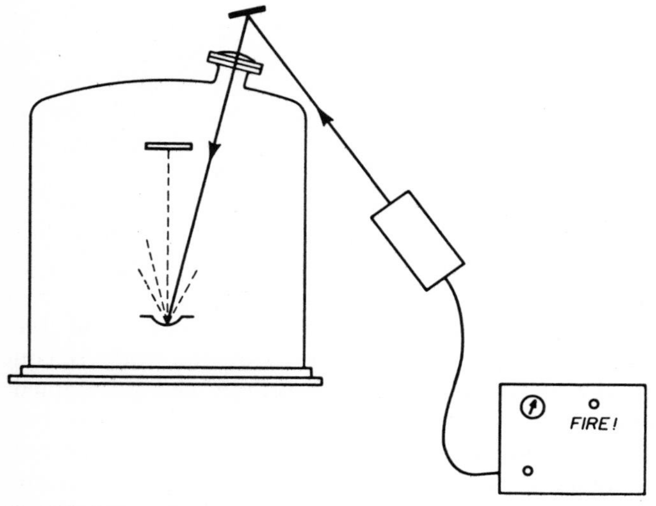

FIG. 2.8. Laser evaporator system.

With the help of focussed electron beams, from one or another of various designs of gun, practically any known material may be evaporated. Thus films of aluminium oxide, or of the somewhat refractory rare earth oxides, are readily made in this way. With the development of the high power laser, an interesting possibility has arisen – of evaporating a material in a vacuum with the whole heating assembly outside, and by a method which involves no crucible heating (fig. 2.8). This method is in a very early stage of development.

We can perhaps distinguish two general areas of thin film production. For the most basic studies of film properties, ultra-high vacuum systems are used in which under favourable conditions the pressure may be kept below, say, 10^{-9} torr during the evaporation, although usually the situation is less happy than this. Nevertheless conditions can be established now such that the rate R at which the substrate surface is peppered with atoms of residual gas, viz:

$$R = \frac{p}{\sqrt{2\pi mkT}} \qquad 2.7$$

is so small compared with that for the condensing atoms that the effects of the residual gas on film properties may be safely ignored.

In the technological area, in which film systems are fabricated for commercial devices, the price paid for the often highly complex arrangements of multiple evaporation sources, shutters and an often bewildering array of internal hardware is that more modest vacua must be tolerated. Nevertheless many quite complicated commercial systems work at pressures around 10^{-7} torr, resulting in a reasonable degree of reproducibility of the systems made.

The relation between the composition of an alloy source and that of the film produced is a complex one and cannot usually be predicted with any certainty. Since one rarely has a condition of equilibrium between the (liquid) source and the vapour stream, then information on the vapour pressures of the constituents of the alloy is usually singularly irrelevant. In some cases, the lower boiling point constituent may come off almost exclusively whereas in others a film of almost the same composition as that of the source may be obtained. Unlike the potential offender against justice, the alloy film should be assumed to be guilty of a composition deviation unless proved innocent. (The latter may be not too easy an exercise, but can be tackled by, e.g. radio-tracer methods, X-ray fluorescence, electron diffraction or even – with some immoderate finesse – by microchemistry.)

2.5 *Comments*

It will be clear on reflection that the above methods of forming films produce their results under a variety of very different conditions. We can adduce many reasons why these processes should *not* produce the theoretician's delight of a parallel-sided slab containing a pristine sample of a 'perfect' crystal. We have mentioned the possible effects of the state of the substrate surface on film properties. We can readily understand that, apart from these uncertainties, others will arise from, say, the effects of carrier gas in sputtered films; of residual gas in the not-so-perfect vacuum-evaporated films; of solute molecules in electrolytic films; of unwanted reactant molecules in vapour-deposited films and in films produced by diffusion processes. There is evidence that the presence of electrons or ions in the condensing beam may influence structure in a marked way. Thus we are not very surprised to observe differences in structure and of physical properties of films produced by different methods (or sometimes, unhappily, of films produced under apparently identical conditions). Rarely is comprehensive information available to enable us to make detailed comparisons between different film systems. We shall, however, try to indicate, in the succeeding chapters, what are the main differences between various film systems. We shall first, however, devote some space to a survey of methods which enable us to gain information on some of the basic properties of films.

3

Methods of Examination of Films

3.1 *General methods*

We may make a broad division of the methods for studying films into categories of the macroscopic and microscopic. In the class of macroscopic methods, we may include (*i*) adsorption methods, such as those traditionally used by the physical chemist in the study of surfaces, (*ii*) X-ray diffraction methods which may be used for thick films and also for single-crystal thin layers, (*iii*) ellipsometry, which enables the optical parameters characterizing a film to be accurately determined. In addition, we may in some circumstances draw valuable inferences from measurements of other physical properties of the films – electrical, magnetic, optical, mechanical – about the type of film structure present. For the most part the above methods yield information on a macroscopic scale. Fine-scale information on films is to be had by the use of the electron microscope, by high- and low-energy electron diffraction methods and, for a limited class of materials, by field-ion microscopy. So far as electron diffraction is concerned, there is a further division inasmuch as the high-energy method gives 'bulk' information about the average structure over depths of many hundreds or thousands of lattice constants whereas low-energy diffraction gives surface information, to depths of at most a few atomic layers.

The strikingly rapid development of the electron microscope has enabled great progress to be made in the study of films. Some of the more important aspects of this work are discussed below.

3.2 *Surface studies of films by adsorption methods*

In many materials in film form, particularly at small thicknesses, the true surface area of solid/outside world interface is considerably larger than the apparent, macroscopic surface area. The factor by which the true area exceeds the apparent has often proved a convenient parameter with which to bend experimental results to fit theory. Direct, completely reliable estimation of this factor is difficult but can, with some reasonable assumptions, be arrived at by the study of adsorption isotherms. A known volume and pressure of gas is connected with an initially-evacuated volume containing the surface under investigation. From the pressure observed after mixing, the number of gas molecules which have been adsorbed on to the test surface may be directly deduced. From the observed relationship between the number of molecules adsorbed and the mass of film, it is sometimes possible to obtain information on the type of film structure present. Thus in the case of measurements on the adsorption of ethane on thermally-evaporated copper films, it has been observed that the number of molecules adsorbed increases linearly with the film mass, which is strongly suggestive of an extremely porous structure.

It seems likely that for certain types of experiment, the low-energy electron diffraction method discussed below may prove an alternative method for studies of this type. However, it is unlikely to be useful for the study of polycrystalline films.

3.3 *X-ray diffraction methods*

On account of the large penetration depth of X-rays, diffraction methods are generally restricted to studies of bulk materials. However, with some precautions, the method may be employed for film studies and is able to yield information on strain and on particle size. It is of particular interest for cases where films must be studied on a bulk surface under conditions where other methods, such as reflection electron diffraction, cannot be applied.

Since the total X-ray scattering by a film of the order of tens of nm thickness is inevitably small, it is necessary that the X-ray source produce minimal background radiation. Use is made for this purpose of a Guinier crystal, typically consisting of a very thin (tenths millimeter) crystal of LiF, bent to the shape of a spherical surface and set up so that all parts of the plate are in the Bragg position with respect to the X-ray tube. The diffracted beam is then used as a source for the film study and the film diffractions are detected by means of a scintillation counter.

Measurements of the directions of the diffracted beam serve to indicate whether the lattice dimensions of the film are the same as those of the substrate, whilst the studies of the line-shape enable information on crystal size and shape to be obtained. Thus fig. 3.1

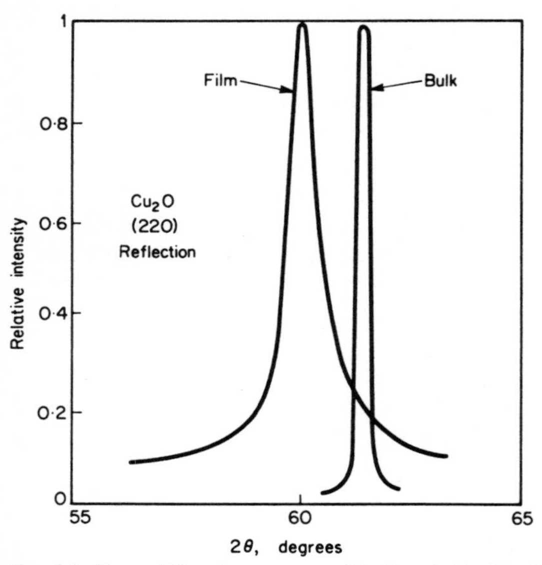

FIG. 3.1. X-ray diffraction patterns of bulk and thin-film Cu_2O.

compares the diffraction patterns obtained from bulk Cu_2O and that of a Cu_2O film 28 nm thick, showing a significantly smaller lattice constant for the thin film material. Before one can happily conclude that this effect is real, one needs to make integrated intensity measurements for a number of diffracted beams in order to determine whether the structure is in fact that of Cu_2O. When this is done, the question is resolved – the structure is the same as the bulk and the unit cell volume agrees closely with the bulk value. The difference in lattice parameters arises from distortion of the lattice, due to strain. Significantly, it is found that the lattice parameters increase with film thickness, suggesting a reduction in strain as the thickness increases.

For thick films – say in excess of one micron – X-ray rocking curves may be used to give a measure of crystalline perfection. The arrangement is shown schematically in fig. 3.2; a Guinier crystal

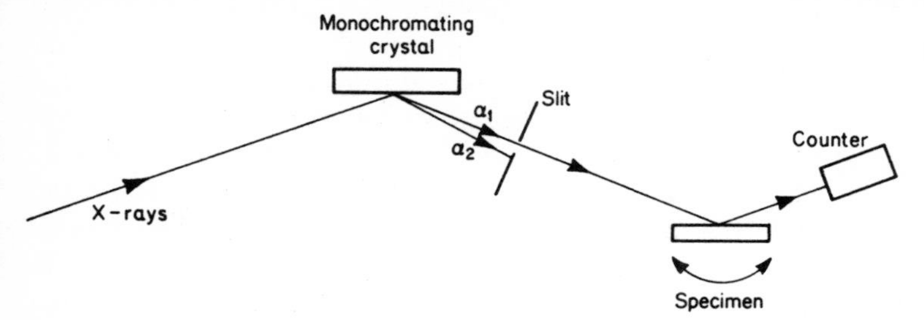

FIG. 3.2. Arrangement for obtaining X-ray rocking curves.

serves as X-ray monochromator, the reflected beam being directed on to the film under test. The film is rocked about a small angle in the neighbourhood of a reflecting position and the reflected intensity monitored by a counter. The more perfect the orientation of the film, the smaller is the angular range over which a diffracted beam emerges.

3.4 *Electron microscopy*

The electron microscope now forms a standard technique for the examination of films and a detailed discussion of the instrument would be out of place in a book of this kind. The instrument and its application are described in Thomas's 'Transmission Electron Microscopy of Metals' (Wiley, 1962). Since the publication of this work, the performance of the best microscopes has reached the stage where individual planes of atoms in metal crystals (e.g. the 200 spacing of gold) can be resolved.

When the structure of films is examined in the electron microscope, variations of brightness and interference fringe effects are often observed in the images produced. In some cases, these effects may be accounted for on kinematical considerations, in which it is assumed that multiple scattering of electrons does not occur, that absorption can be neglected and that the monoenergetic beam incident on the specimen is *not* along a direction which satisfied the Bragg condition for any planes in the crystals. The kinematical theory treats the atoms in the crystal as (point) scattering centres, with a scattering power depending on the atomic number of the atom.

Although in some circumstances kinematical theory gives a fair description of the experimental observations, there are many situations in which it is necessary to take account of the reality that the crystal possesses a periodic lattice potential, with which the electron wave interacts. It must also be acknowledged that the fraction of an incident electron beam which is diffracted may be quite large and that multiple scattering may play an important role. The general approach to this problem, known as dynamical theory, leads to equations which cannot be solved analytically except for rather simplified situations. Nevertheless the forms of solution accord well with the experimental results in many cases and show significant differences from the predictions of kinematical theory.

The dependence of the intensity of an electron wave scattered in a particular direction on the unit cell parameters and the crystal thickness may be readily calculated by kinematical theory and takes a simple form for the case of diffraction through a small angle – a situation generally met in practice. For an electron beam in a direction $\mathbf{S}_0$ and a pair of atoms (fig. 3.3) at O and P (where $OP \equiv \mathbf{r}_i$) the phase difference between the waves scattered in a

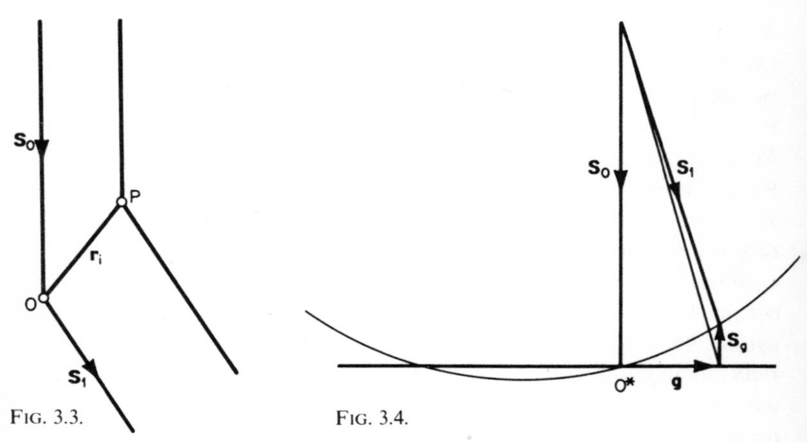

FIG. 3.3.

FIG. 3.4.

direction $\mathbf{S}_1$ is simply $(2\pi/\lambda)(\mathbf{S}_1 - \mathbf{S}_0).\mathbf{r}_i$. If the reciprocal lattice vector of the diffracting planes is $\mathbf{g}$, then the diffracted intensity, for the small-angle diffraction case, may be simply expressed in terms of the crystal thickness l and the distance, $\mathbf{S}_g$, in reciprocal space of the reciprocal lattice point from the Ewald sphere, as shown in fig. 3.4. The diffracted intensity is simply proportional to

$$\sin^2(\pi l \mathbf{S}_g)/(\pi \mathbf{S}_g)^2.$$

Since the intensity varies sinusoidally with l, then interference fringe effects are to be expected. In particular, if a parallel-sided film contains a stacking fault running obliquely across it, then the emerging intensity will oscillate as we pass across the fault region, depending on the relative phases of the diffracted waves from the regions above and below the fault. Thus such stacking faults may immediately be recognized from the fringes observed in a micrograph of the specimen.

The case of a crystal containing imperfections can be fairly easily dealt with on kinematical theory. If an atom which in the perfect crystal lies at $\mathbf{r}_i$ is displaced to $\mathbf{r}_i + \Delta\mathbf{r}_i$, then the diffracted amplitude is given by

$$A = \int \exp 2\pi i\{\mathbf{g}.\Delta\mathbf{r}_i + \mathbf{S}_g.\mathbf{r}_i\}\mathrm{d}r. \qquad 3.1$$

The value of $\Delta\mathbf{r}_i$ will depend on the type of fault.

For a perfect crystal, the expression given above for the intensity of a diffracted beam displays a periodicity with $(l\mathbf{S}_g)$. If the crystal is set with respect to the beam so that the reciprocal lattice point for the diffracted beam lies *on* the Ewald sphere, then $\mathbf{S}_g = 0$ and no periodicity in the diffracted intensity with crystal thickness would be expected. In fact such a periodicity is found experimentally even for this condition. This illustrates a limitation of kinematical theory. Effects such as these are correctly predicted by dynamical theory. If, however, the crystal is not aligned closely to the position of strong diffraction, then the kinematical and dynamical theories give closely similar results. This is to be expected since for this condition, the diffracted intensity is low so that the effects of multiple scattering are not of great importance.

When films are grown epitaxially on single crystal substrates which are sufficiently thin for electron microscopy, interference fringes are observed. These arise from the Moiré effect of two gratings and can provide invaluable information about the individual crystal lattices in cases where the microscope has insufficient resolving power to enable the lattices to be studied directly. Thus if two gratings of spacing d_1 and d_2 (where $d_1 > d_2$) are superposed with their rulings parallel to one another, Moiré fringes are observed with a spacing $d_1 d_2/(d_1 - d_2)$. Thus if a gold crystal (lattice spacing 0·407 nm) lies in parallel orientation on a copper crystal (spacing 0·361 nm), then the separation of the Moiré fringes produced is 3·19 nm, which can be easily resolved by the normal microscope. The powerful feature of this type of pattern is that the Moiré fringes will register the presence of irregularities in one or other of the lattices. Thus if one of the films contains an edge dislocation, the resulting Moiré pattern shows an identical structure (Plate I).

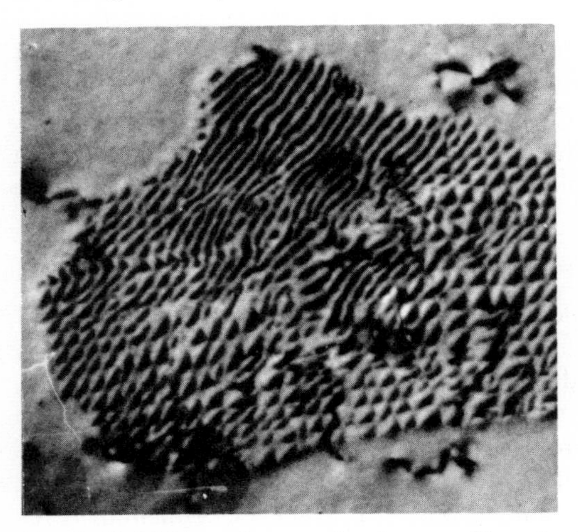

Plate I. Moiré fringes from overlapping crystallites of gold and copper.

If the two lattices which form the Moiré pattern are rotated with respect to each other, then the Moiré fringes also rotate. This provides a rather sensitive measure of the orientation of crystallites one on another, since the angle of rotation of the Moiré pattern is generally much larger than that between the lattices. When such fringes are observed in a growing film, they are often seen to change direction abruptly, indicating a sudden rotation of a whole crystallite.

3.5 *High-energy electron diffraction*

This method gives valuable direct information on the arrangement of atoms in the crystallites of the film and of their relative orientations. It further enables certain structural features, such as twinning, stacking faults and superlattice formation to be directly observed.

When a beam of electrons passes through a solid film, a number of possible interactions can occur. Energy may be lost through processes such as ionisation, secondary emission, excitation or through the generation of X-rays. In addition, electrons may be scattered without loss of energy by the positively charged nuclei of the atoms, a process termed coherent scattering. If the atoms are arranged in a regular array, as in a crystal, then for certain directions, the coherent scattering may assume large values. The scattering $f(\theta)$ for a direction θ (deviation 2θ) is related to the atomic number Z and electron wavelength λ by

$$f(\theta) = \frac{me^2}{8\pi^2\hbar^2}\frac{(Z-f)}{(\sin\theta/\lambda)^2} \qquad 3.2$$

where f is the scattering factor for X-rays. The scattering power of solids for electrons in this process exceeds that for X-rays by some four orders of magnitude. Thus electron diffraction is a useful method for thin layers and cannot, in the velocity range mentioned above, be used for specimens much thicker than a few hundred nm.

The de Broglie wavelength associated with an electron of relativistic mass m, rest mass m_0 and energy E is

$$\lambda = \frac{2\pi\hbar}{\sqrt{(m+m_0)eE}}. \qquad 3.3$$

High-energy electron diffraction is generally carried out using electrons accelerated by potentials of 50–100 kV so that the wavelengths involved are in the neighbourhood of 3–5 pm. The relation of these figures to the spacings of planes in crystals is that diffraction angles of the order of a degree are observed, leading to well-resolved diffraction patterns with a specimen-to-plate distance of the order of half a metre or so.

The directions of the diffracted beams are related to that of the incident beam by the need to fulfil the condition that scattering from corresponding atoms in the crystal shall reinforce. If a, b and c are the repeat distances in the direction of the crystal axes and $(\alpha_0, \beta_0, \gamma_0)$ the direction cosines of the incident beam, then a maximum will

Plate II. Diffraction pattern of epitaxial (100) film of nickel.

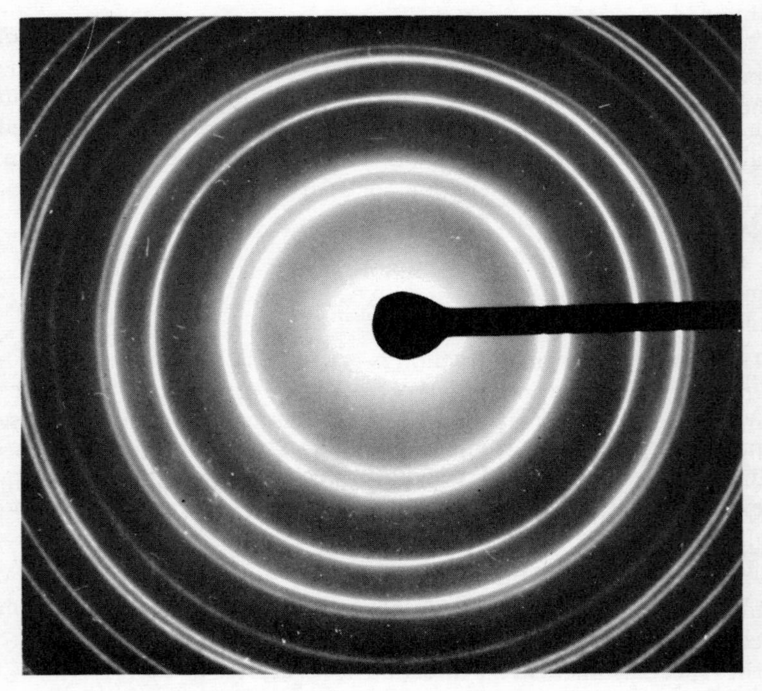

Plate III. Diffraction pattern of polycrystalline nickel film.

occur in the direction (α, β, γ) provided

$$\begin{aligned} a(\cos\alpha - \cos\alpha_0) &= h\lambda \\ b(\cos\beta - \cos\beta_0) &= k\lambda \\ c(\cos\gamma - \cos\gamma_0) &= l\lambda \end{aligned}$$

where h, k, l are integers. If d_{hkl} is the separation between lattice planes with Miller indices (h, k, l) then the foregoing equations imply that

$$2d_{hkl}\sin\theta = \lambda \qquad 3.4$$

where 2θ is the deviation of the beam in the diffraction process. The intensities of the diffracted spots depend on the scattering powers of the atoms and on the atomic arrangement. Thus for face-centred-cubic materials, diffracted spots occur only when (h, k, l) are all odd or all even. For body-centred-cubic materials, $h+k+l$ must be even. For typical high-energy diffraction set-ups, the diffracting pattern obtained represents directly a section through the reciprocal lattice of the crystal concerned. Thus Plate II shows the pattern for a nickel specimen, taken with the beam perpendicular to the (100) plane of the specimen. When a polycrystalline specimen is examined, the resultant pattern is a superposition of large numbers of patterns such as that of Plate II, with all possible orientations around the beam directions, leading to the ring pattern of Plate III. In certain cases, partial orientation occurs, leading to patterns like that of Plate IV.

If the specimen under examination is a good single crystal, Kikuchi lines (Plate V) may be observed. These arise from the diffraction of diffusely scattered electrons, as illustrated in fig. 3.5.

In the absence of diffraction effects, the angular distribution of diffusely scattered electrons would be of the form shown in fig. 3.6. Diffraction at the set of planes shown results in an excess of electrons travelling in the ϕ_1 direction and a deficit in the ϕ_2 direction. These directions form cones around the normal to the planes and thus intersect a flat plate placed perpendicular to the beam in conics – which in practice emerge as practically straight lines. Clearly such effects would be seriously obscured if different parts of the 'single' crystal were even slightly misaligned. The occurrence of clear Kikuchi patterns is a strong indication of crystal perfection.

Certain types of crystalline imperfection are immediately detectable in electron diffraction patterns. One of the commonest observed

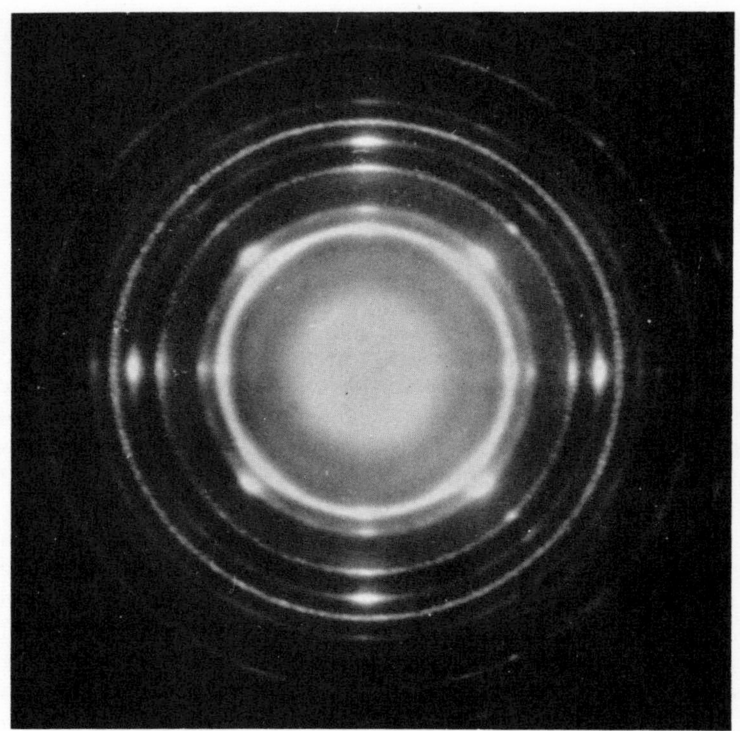

Plate IV. Partial orientation in nickel film.

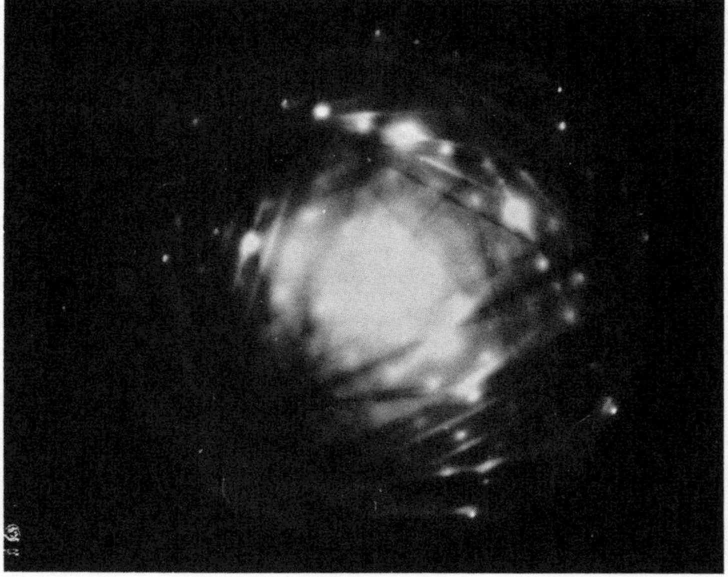

Plate V. Kikuchi lines from epitaxial nickel film.

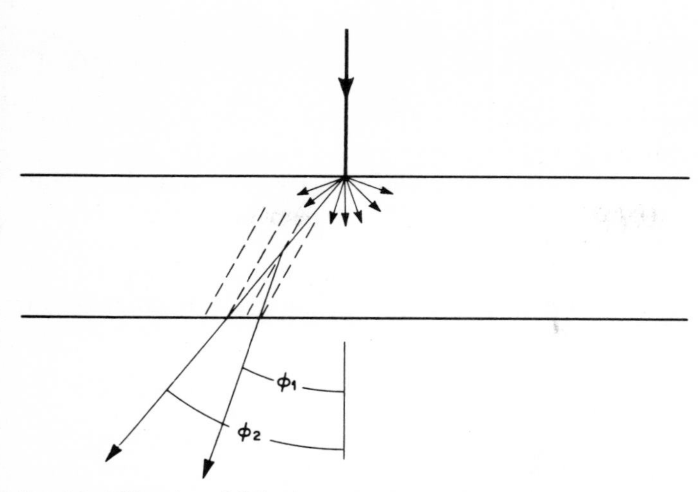

FIG. 3.5. Diffraction of diffusely scattered electrons.

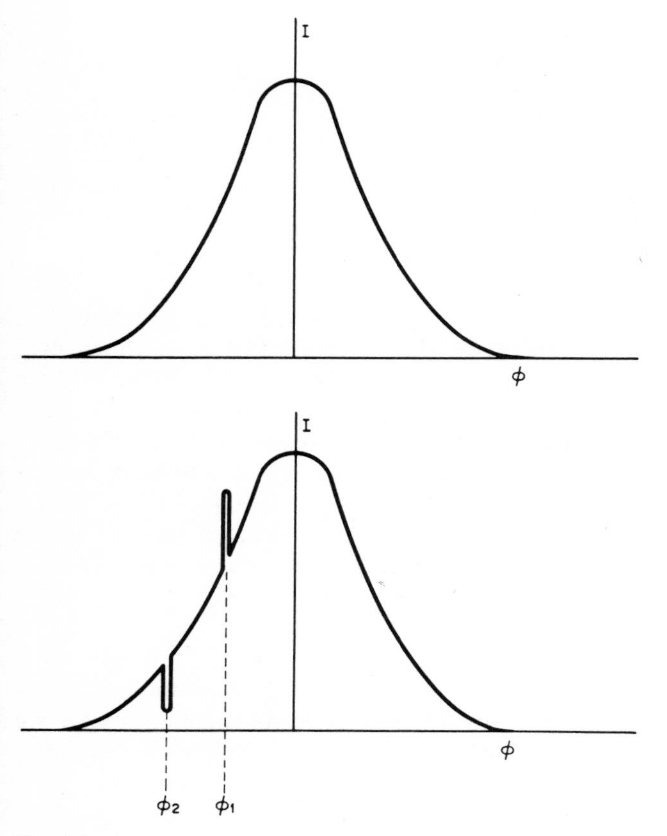

FIG. 3.6.

in specimens of face-centred-cubic materials is the stacking fault. If we consider the assembly of a crystal by laying down (111) planes we note that, on characterizing the first layer by A and the next by B, we have a choice of two possible positions for the third layer. If it is placed above the atoms in the A layer, and the sequence ABAB . . . repeated, a hexagonal structure results. Placing the third layer (fig. 3.7) in the alternative (C) position and building up the

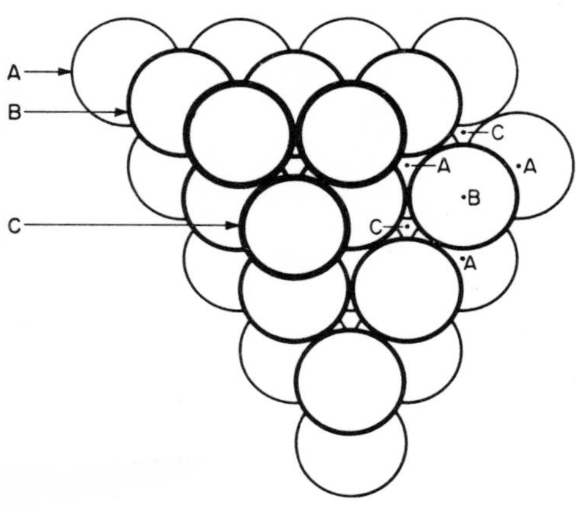

FIG. 3.7. Alternative stacking of (111) type planes.

sequence ABCABC . . ., yields a face-centred-cubic structure. The occurrence of a sequence ABCABCABABCABC . . ., which may be imagined as due to a missing C plane, or to a shear in the <112> direction constitutes a stacking fault. Extra streaks or spots occur in the diffraction pattern. The specimen may be thought of as containing extremely thin hexagonal regions, for which the reciprocal lattice spots will be sufficiently elongated to intersect the Ewald sphere, giving rise to additional spots. A detailed analysis shows that the spots of the f.c.c. lattice which are affected by the presence of stacking faults on (111) planes are those for which $h+k+l = 3n+1$ where n is an integer.

Another example of the way in which extra diffraction spots indicate structure features is in the case of the formation of superlattices in the specimen. In the classic case of the alloy CuAu, we have a high temperature form in which lattice sites on a face-centred-cubic array are occupied randomly by Au or Cu atoms. Thus the normal pattern (fig. 3.8) for a face-centred-cubic crystal is obtained. At low temperatures, an ordered structure shown in fig. 3.9 is formed,

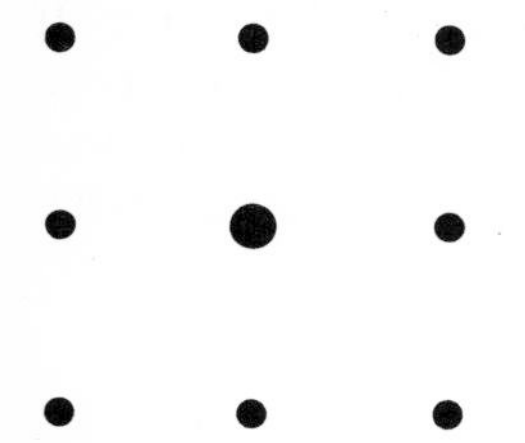

FIG. 3.8. Form of diffraction pattern for disordered CuAu structure.

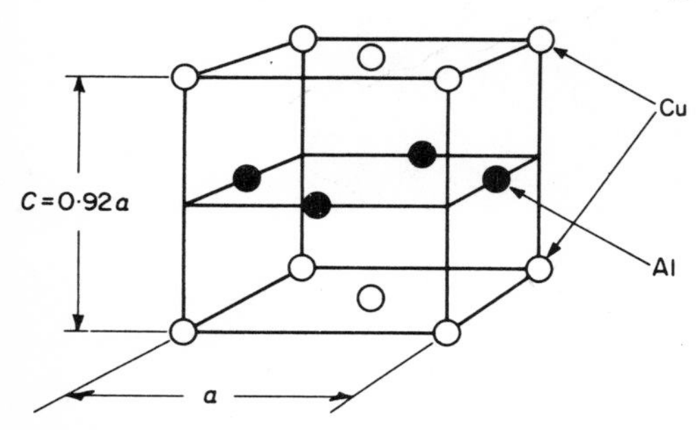

FIG. 3.9. Tetragonal structure of ordered alloy.

in which alternate (100) planes contain all Au and all Cu. The unit cell is now tetragonal, with a c/a ratio of 0·92. At intermediate temperatures – and for no known reason – a structure with a periodicity of about 2 nm forms, as indicated in fig. 3.10. The long

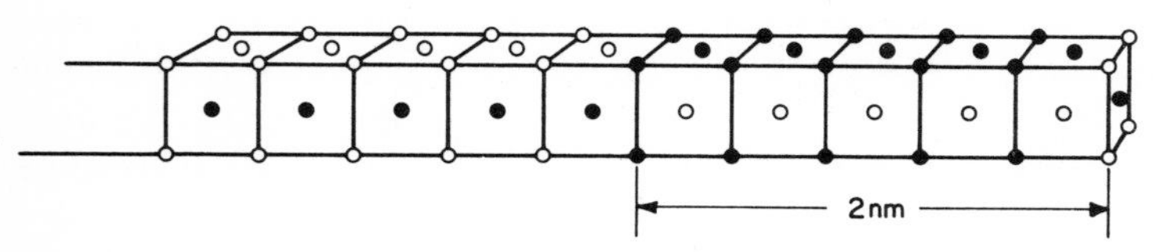

FIG. 3.10. Unusual structure of CuAu (CuAu II).

periodicity is manifested by the occurrence of multiple spots (Plate VI) in the neighbourhood of the (100) type spots, which are forbidden diffractions in the disordered structure. When at low temperatures, the pattern of Plate VI changes to that of fig. 3.8, the quartets of spots coalescing into single ones.

Plate VI. Diffraction pattern of CuAu II. Structure as shown in fig. 3.10.

To examples such as those given above may be added the many straightforward crystal growth effects which may be studied by high-energy electron diffraction–twinning, double diffraction, revelation of atomic flatness by refraction effects in reflection patterns. Many of the features of high-energy diffraction patterns may be fairly fully accounted for provided the appropriate theory, as discussed below, is applied. There remain other features, e.g. the very low intensity diffuse streaks which join the spots of apparently good crystal patterns, for which no explanation is yet available.

The detailed discussion of the intensity distribution in the diffracted spots depends on the thickness of the specimen being examined. If the specimen is thin enough for there to be negligible rescattering or absorption, and provided the diffracted intensity is very small compared with that of the incident beam, the kinematic theory of diffraction applies. This treats the crystal as a set of point scattering centres, taking no account of the actual potential distribution in the crystal or of the effects of crystal boundaries. The scattering intensity per unit area of specimen for which the volume of the unit cell is V is given, for a Bragg angle θ, by

$$I(\theta) = \frac{\lambda^2 f^2 \sec^2\theta}{V^2} \frac{\sin^2(\pi ts)}{(\pi ts)^2} \qquad 3.5$$

where t is the specimen thickness and s is the distance from the reciprocal lattice point to the Ewald sphere measured in the direction normal to the specimen plane.

When the restricting conditions of the kinematic theory are not met, it is necessary to use the dynamical theory, for which the starting point is the (non-relativistic) Schrodinger equation. For the potential function in this equation, the electron density in the crystal needs to be represented by a three-dimensional Fourier series. In solving the equation to zero order of approximation, the result is precisely that of the kinematic theory. As solutions are taken to higher orders, so more fine detail appears in the solution which then represents, in many cases, the kind of detail observed in experimental plates.

The importance of the dynamical theory of diffraction is evident when, in electron micrographs, extinctions and fringe effects are seen. In order that misinterpretations of such effects may be avoided, a clear understanding of the mechanism of diffraction effects is essential.

3.6 *Low-energy electron diffraction*

Although electrons with energies in the many kilovolt region may be used to study surfaces by reflection electron diffraction, the fact that such electrons penetrate to depths of the order of tens of nanometres makes the method of limited use for examination of the atomic layers close to the surface. Since low-energy electrons – say in the 10 to 200 eV region – penetrate only to depths of the order of a few atomic layers, these form a far more powerful probe. Although electron diffraction was first observed in 1927 by two groups, one using high-energy electrons (Thomson) and the other employing slow electrons (Davisson and Germer), it was the high-energy method which was mainly exploited in the next three decades following the discovery. The reason is simple. The very fact that low-energy electrons penetrate surfaces to such small depths meant that inevitably diffraction effects at surfaces would be extremely susceptible to minute amounts of surface contamination, including the effects of adsorbed gases in the specimen chamber. Since for most of the period under discussion it was somewhat difficult to produce surfaces which remained contamination-free for more than a very short time, the sensitivity of the method proved at the same time a limitation. One flourishing group proved the exception – Farnsworth and his co-workers had, by scrupulous attention to detail and by exceptional experimental skill, been able to make valuable studies using electrons with energies in the range up to a few hundred eV. In these experiments, the diffracted electrons were collected in a Faraday cage which was traversed over the various required angular ranges and the collected current was measured. On account of the very low currents involved, this procedure was inevitably slow so that the amount of information which could be obtained from an initially clean surface was necessarily limited (Schlier and Farnsworth, 1957).

After a long period away from the field of electron diffraction, Germer returned to the fray in 1959 and, by implementing a suggestion made by Ehrenberg *25 years* earlier, abruptly brought the technique of low-energy electron diffraction to the status of a standard tool, such as had been the case for fast electrons.

The step needed was such a simple one that it seems strange that it had not been taken before, especially after it had been suggested. The point with slow electrons is that they have insufficient energy to produce fluorescence on the kind of screen used for fast electrons. Ehrenberg's suggestion was elegant in its simplicity: if the electrons are not going fast enough to produce any scintillation, accelerate them until they are! (The technique of post-deflection acceleration had in fact been used in cathode-ray tube technology.)

In fact, a strong contributory reason for the rapid growth of low-energy diffraction is that at around this time, the business of producing vacua in the region of $10^{-9} - 10^{-10}$ torr was rapidly moving from the specialized laboratory and was becoming a routine facility. Nevertheless the feasibility of obtaining a complete diffraction record in a matter of seconds instead of hours constituted a very significant advance.

There is one further essential refinement in dealing with slow electrons. Whereas for high-energy electrons, practically the whole of the transmitted (or reflected) beam arises from coherent scattering, with negligible energy loss, in the case of slow electrons, the extent of coherent scattering may amount to only about 1% of the incident beam. If all the scattered electrons were accelerated on to a screen, the record of coherently scattered electrons would be completely masked by the effects of the remaining 99% of the electrons, all of which would have lost various amounts of energy. Exclusion of these electrons is trivially achieved by means of a suitably biased grid (fig. 3.11).

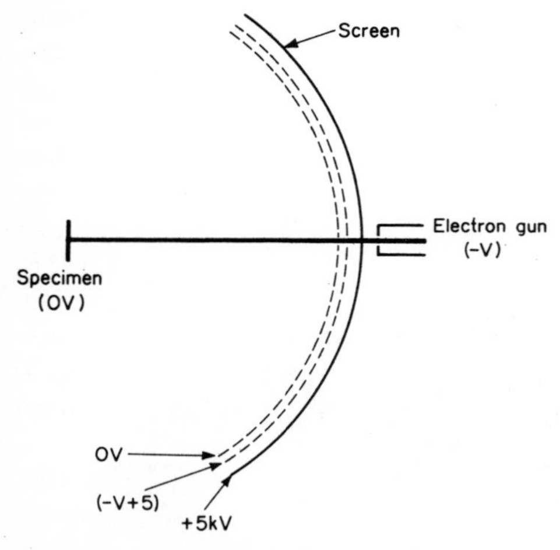

FIG. 3.11. Schematic of LEED system.

As the penetration depth of slow electrons amounts to only a few atomic dimensions, the specimen behaves like an extremely thin crystal. The associated reciprocal lattice points will therefore be very long and the pattern obtained will arise virtually as though due to

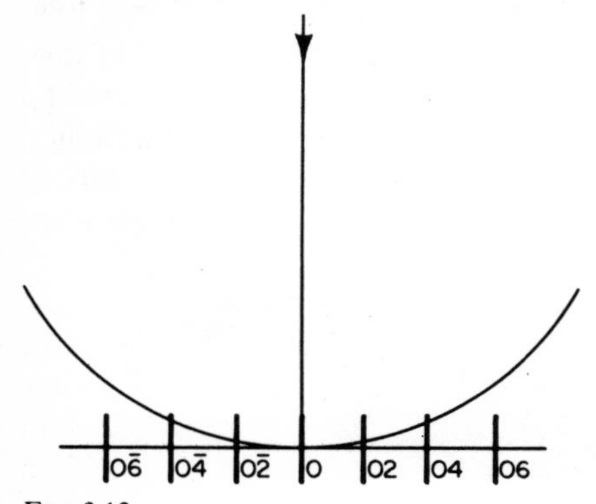

FIG. 3.12.

a two-dimensional grating. Thus fig. 3.12 shows the Ewald sphere for 900 volt electrons (λ = 41 pm) incident normal to a (100) plane of a silver crystal (a = 0·404 nm).

Although the interpretation of the *positions* of diffracted spots in low-energy diffraction patterns is fairly straightforward, the interpretation of the *intensities* of the spots is far more difficult, on a truly quantitative basis. This is because the detailed theory of the interaction of electrons of such energies with solid surfaces is yet to be established.

Notwithstanding this limitation, the use of slow electrons in surface investigations is expected to throw light on some of the basic ideas on the growth of thin films. The phenomenon of epitaxy

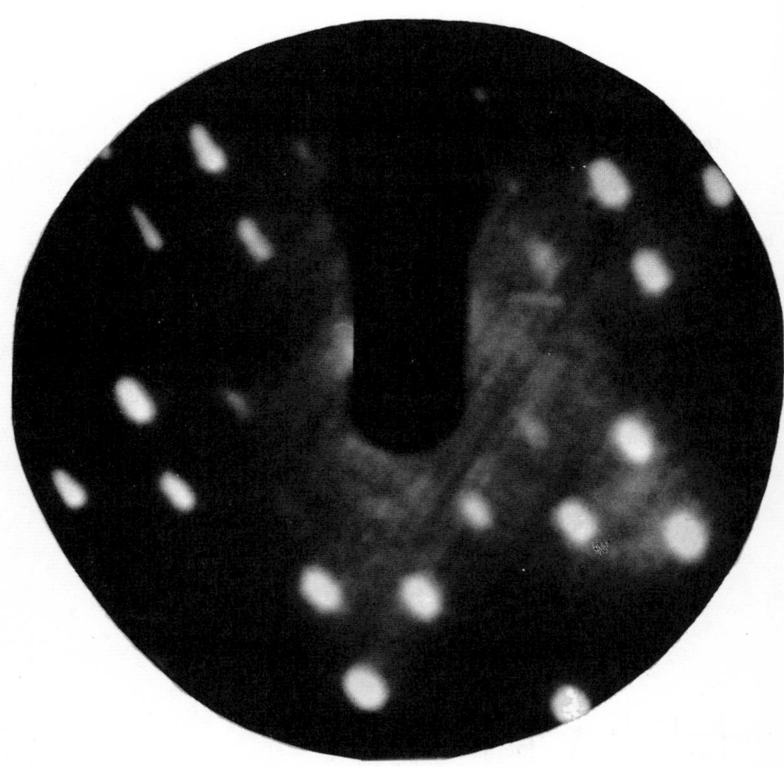

Plate VII. LEED pattern of nickel surface partly covered by carbon monoxide.

remains almost as deep a mystery now as when it was first observed. Single-crystal overgrowth occurs generally only when the substrate is itself monocrystalline *but the relative values of lattice spacings in substrate and deposit seem totally unimportant*. It is still not obvious why this should be so but it is surely true that the behaviour of the first few atoms to arrive on the surface must be of great importance. The technique of low-energy electron diffraction provides for the first time a tool of sufficient sensitivity to enable questions about such initial layers to be answered.

An indication of the power of the method may be seen from Plate VII which shows the diffraction pattern from a nickel surface covered by a quarter of a monolayer of carbon monoxide. The inner spots in the pattern arise entirely from the adsorbed CO molecules. The distribution of CO molecules on the surface is a rectangular one, shown in fig. 3.13. Since the nickel surface (which is an octahedral face) has threefold symmetry, three orientations of the rectangular array are possible, inclined at 120° to one another. The threefold symmetry of the diffraction pattern shows that all three orientations are present. The pattern is described as a '$2 \times \sqrt{3}$' distribution, the logic of which follows from fig. 3.13.

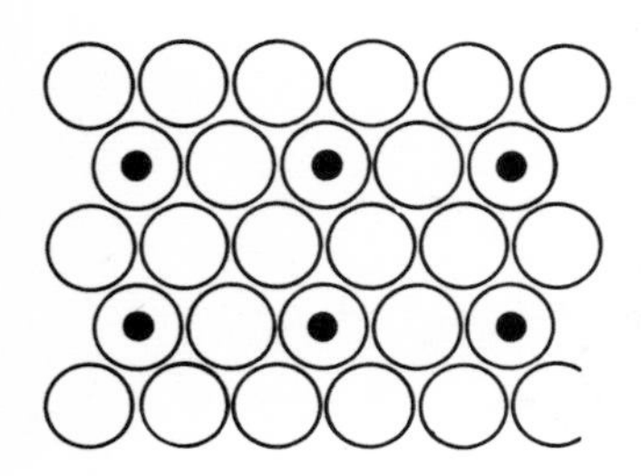

FIG. 3.13. Distribution of carbon monoxide on nickel surface, deduced from pattern of Plate VII. The adsorbed CO occupies one quarter of the surface sites. Three orientations occur, inclined at 120° to one another.

3.7 *Auger emission spectroscopy*

A powerful newcomer to the battery of techniques for studying thin films and surfaces is that of Auger emission spectroscopy. Whilst diffraction of low-energy electrons gives information on the arrangement of atoms on a surface, it is not always possible unequivocally to identify the actual atomic species on the surface. This may become possible when the theory is better developed, but there will always arise the difficulty that atoms of approximately the same mass will scatter nearly equally and so will be difficult to distinguish.

When an incident electron displaces an inner shell electron from an atom, then an electron from a higher energy shell makes a transition to the lower shell and an X-ray is emitted. In some cases, this X-ray emerges from the atom and can be detected. This is the well-established process of X-ray spectroscopy and is a useful tool for studying heavy elements. For lighter elements it is not unusual for the X-ray quantum to be absorbed within the emitting atom and for an electron (Auger electron) from one of the other shells to be ejected. In this case the emitted electron leaves with an energy characteristic of the emitting atom. If then the energy spectrum of the electrons emitted by the atom is measured, a peak is found at this characteristic energy. It is usual to display the derivative of the energy curve, rather than the distribution itself, the conversion being easily effected electronically.

For a surface of which a fraction θ is covered by the atoms to be identified, and for a primary beam current of I μA, the order of magnitude of the Auger electron current is $\sim 10^{-11}\theta$ amperes. With appropriately sensitive detection arrangements, it is possible to identify $\sim$0·1 monolayer on a surface. A typical Auger curve, for nickel on potassium iodide, is shown in fig. 3.14.

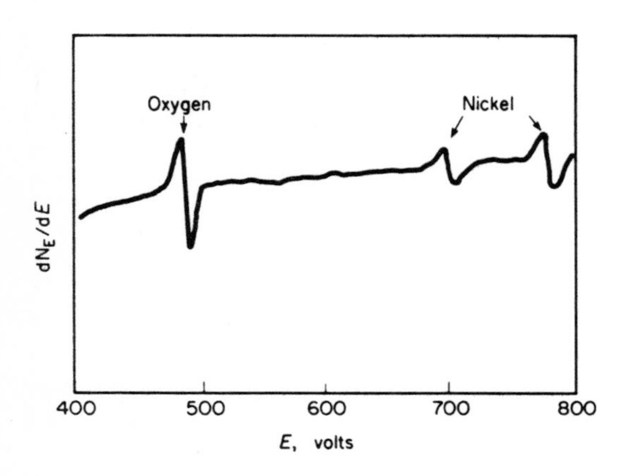

FIG. 3.14. Spectrum of Auger electrons emitted from potassium iodide surface covered with nickel.

4

Growth and Structure of Films

4.1 *General comments*

The structure exhibited by materials in thin film form depends on an array of factors so depressingly large that only rather broad generalizations are possible. In the case of films deposited by thermal evaporation, the more important of such factors are:

Pressure and nature of residual gas in deposition chamber;
Temperature of evaporation source;
Rate of deposition of condensing atoms;
Temperature of substrate;
Surface mobility of deposit atoms on substrate;
Nature of substrate (amorphous, polycrystalline or monocrystalline);
Presence of charged particles (ions, electrons) either in, or in the neighbourhood of, the depositing atoms;
Presence of electric or magnetic fields at the substrate surface;
Occurrence of chemical reactions between deposit and substrate.

In the early days of film studies, the influence of some of the above factors was not realized and the techniques needed for controlled, definitive experiments were not available. This led inevitably to a high degree of speculation on the nature of the processes occurring when atoms deposit on a surface to form a film. Although rapid progress has been made in recent years, we are still some way from a detailed understanding of the processes of nucleation and growth. An exhaustive study of this aspect of film formation will therefore not be made, but rather some general indications of the problems involved. On the theoretical side, there are two approaches to the problem of film nucleation. One is an atomistic approach, which aims at discussing the formation of nuclei in terms of adsorption and diffusion energies of deposit atoms on the substrate and in terms of self-adsorption. Although this is a potentially powerful approach, the problem arises that the necessary parameters are rarely known and are often very difficult to measure. One exception to this is the case of condensed rare gas atoms, for which such data are available. The other main theoretical approach is the capillarity theory, which treats the problem in terms of macroscopic concepts such as surface energy and contact angles. There is evidence that for treating the growth of large crystallites, such ideas are useful. By analogy with the behaviour of liquid/vapour systems, it would be expected that for given deposition conditions, a critical size of nucleus will exist above which continuous growth will occur. One of the difficulties of capillarity theory is that the critical size appears in some cases to be of the order of a fraction of a nm, so that such a critical nucleus would consist only of a small number of atoms. In such situations, the use of macroscopic thermodynamic concepts is clearly of doubtful validity. One further crucial feature which has been revealed by cinemicroscopy studies in the electron microscope and which cannot easily be embodied in theories of growth, is the high mobility of relatively massive nuclei on a surface, even at temperatures very far below the melting point of the deposit material.

The powerful array of techniques which have become available in recent years and which are described in Chapter 3 are such that many of the early uncertainties are now being resolved. However, although the many techniques separately provide invaluable information, it is not possible to employ them all simultaneously to a single situation. Thus, whilst low-energy electron diffraction and Auger spectroscopy can be performed on a single system, it is not at present possible to examine the macroscopic structure simultaneously by conventional electron microscopy. Also much of the information gained by electron microscopy has perforce, owing to the difficulties involved in operating such an instrument under ultra-high vacuum, been obtained under relatively poor vacuum conditions.

Uncertainties such as the above make it difficult to be sure that the conditions underlying the assumptions of any theoretical treatments have been met in experimental studies. This is particularly true of attempts to verify theories of nucleation. Since reliable values for binding energies, activation energies for diffusion and surface mobilities are only rarely available, detailed verification of theory cannot at this stage be expected. In certain cases, the interdependence of some of the observed deposition variables is found to follow theoretical predictions for plausible values of the energy parameters involved.

4.2 *General features of nucleation theories*

The underlying basis of theories of nucleation derives from the work of Gibbs who considered the stability of a spherical cluster of atoms of material in one phase within a homogeneous region of a second phase. Such considerations apply to the formation of droplets from a vapour and, in the case of large nuclei, to the case of vapour → solid condensation. Further development is required, as shown below, to deal with the case of nucleation on a surface which may be of different material. For the formation of a liquid drop from a supersaturated vapour at pressure p, the free energy of formation ΔG_0 of a spherical cluster of radius r is given by

$$\Delta G_0 = 4\pi r^2 \sigma + \frac{4\pi r^3}{3}\left(\frac{kT}{\Omega}\right) \ln (p/p_e) \qquad 4.1$$

where σ is the interfacial energy of the liquid, p_e the equilibrium vapour pressure of the liquid and Ω the molecular volume. ΔG_0 has a maximum value, from equation 4.1, at a value of the drop radius r_m given by:

$$r_m = -\frac{2\sigma\Omega}{kT \ln (p/p_e)}. \qquad 4.2$$

For clusters with greater radius than this value, the free energy decreases with increasing size, so that continued growth of the droplet occurs.

The maximum value of the free energy, ΔG_m, corresponding to the radius r_m is given by

$$\Delta G_m = \frac{16\pi\sigma^3}{3\left[\frac{kT}{\Omega} \ln (p/p_e)\right]^2}. \qquad 4.3$$

If we now consider the dynamic picture of atoms impinging on existing nuclei, being adsorbed and then re-evaporating after a mean-free-time τ, then we may readily obtain an expression for the rate of formation of critical nuclei. For an impingement rate R, the equilibrium concentration of atoms in the nucleus is $n_e = R\tau$. Since the energy of formation of a nucleus of radius r_m is ΔG_m, then the equilibrium concentration n^* of critical nuclei is given by

$$n^* = n_e \exp(-\Delta G_m/kT). \qquad 4.4$$

Now at a vapour pressure p, the number of atoms striking unit area of surface in unit time is $p/(2\pi mkT)^{\frac{1}{2}}$. If all the atoms condense, then this is the rate of increase of the number of surface atoms. In general, we may characterize a real surface by an accommodation coefficient α, the ratio of the number of atoms which 'stick' to the total number incident. The number condensing per unit time on unit area is thus $\alpha p/(2\pi mkT)^{\frac{1}{2}}$ and the number condensing per unit time on the surface of a critical nucleus is given by $\alpha p/(2\pi mkT)^{\frac{1}{2}} \times 4\pi r_m^2$. Thus the nucleation frequency I under these conditions of homogeneous nucleation is given by

$$I = \frac{4\pi r^2 \alpha p}{(2\pi mkT)^{\frac{1}{2}}} \cdot n_e \exp(-\Delta G_m/kT) \qquad 4.5$$

a result revelling in the name of the Volmer-Weber-Becker-Döring equation.

The extension of these ideas to deal with condensation on to a substrate necessitates the inclusion of terms to allow for the effects of desorption and of surface diffusion. The capillarity model of surface nucleation treats the case of a hemispherical cap of condensed material (fig. 4.1). It is assumed that the surface possesses a density

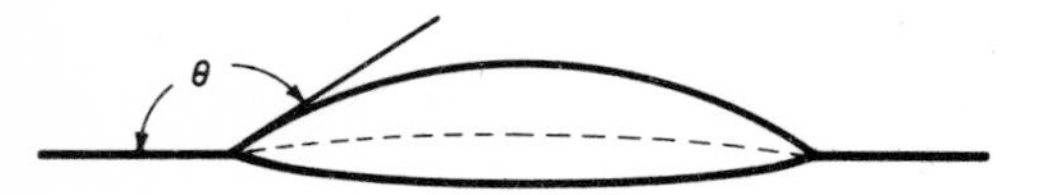

FIG. 4.1. Crystallite shape assumed in capillarity model.

n_0 of adsorption sites, for which the free energy for desorption is ΔG_{des}. Surface diffusion, envisaged as a jump process between adsorption sites, is characterised by a free energy ΔG_{diff}. For this case, the nucleation equation takes the form (Hirth and Pound, 1964)

$$I = \frac{2\pi Z r_{\mathrm{m}} n_0 a p \sin\theta}{(2\pi m k T)^{\frac{1}{2}}} \exp\left[\frac{\Delta G_{\mathrm{des}} - \Delta G_{\mathrm{diff}} - \Delta G'_{\mathrm{m}}}{kT}\right] \quad 4.6$$

where
$$\Delta G'_{\mathrm{m}} = \frac{16\pi\Omega^2\sigma_{\mathrm{cv}}^3\phi(\theta)}{3[kT\ln(p/p_{\mathrm{e}})]^2}.$$

Here σ_{cv} is the interfacial surface energy between the deposit material and vapour, $\phi(\theta)$ is a geometrical factor involving the angle of contact of the deposit on the surface and Z is a correction factor allowing for the departure from equilibrium which occurs among the clusters.

As would be expected, the application of homogeneous nucleation theory is restricted to situations in which the number of atoms in the critical nucleus is large enough for the macroscopic parameters such as interfacial energy and angle of contact to be meaningful. Hirth and Pound suggest that these considerations apply for nuclei containing more than ~ 100 atoms.

In many experimental situations, there is evidence that the critical nuclei may consist of fewer than ten atoms and in this case a theory of heterogeneous nucleation is required. The work of Walton (1962) and Rhodin and Walton (1963) is directed to this end. By treating small clusters of atoms as individual molecules, Rhodin and Walton express the concentration n_i^* of clusters of i atoms in terms of the concentration n_0 of adsorption sites and the potential energy E_{0i} of decomposition of such a cluster at absolute zero in the form

$$\frac{n_i^*}{n_0} = \left(\frac{n_1}{n_0}\right)^{i^*} \exp\left(-\frac{E_{0i^*}}{kT}\right) \quad 4.7$$

where n_1 is the surface density of adatoms. The Rhodin-Walton theory leads to the result that the nucleation rate depends mainly on the factor $\exp[(i^*+1)\Delta G_{\mathrm{des}} - \Delta G_{\mathrm{diff}} + E_{0i^*}/kT]$ which suggests that if, as would be expected, the critical size of nucleus varies with temperature, then this will be manifested in a change in the slope of the curve of $\ln I$ vs $1/T$ for deposition at a constant incidence rate. This is indeed observed in many cases and is illustrated in fig. 4.2. The theory also predicts, through a factor R^{i^*}, where R is the rate of incidence, a dependence of the size of critical nucleus on the rate of arrival of atoms, for a constant surface temperature. These general features are also confirmed by experiment.

From the results of heterogeneous nucleation theory, it seems likely that for some materials and a substrate at low temperature, the critical nucleus may consist of only a single atom. Above a critical temperature, critical clusters may consist of two or more

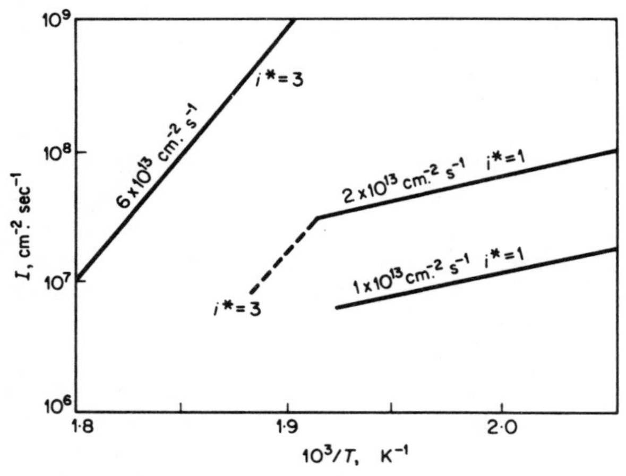

FIG. 4.2. Deposition of silver on the (100) face of sodium chloride, cleaved in ultra-high vacuum.

atoms. Since such clusters have perforce directional characteristics it is tempting to relate the critical temperature to an 'epitaxial' temperature. It is certainly true that in many cases epitaxy occurs only above a particular temperature, although there is evidence that this temperature may depend on factors other than the substrate/deposit materials. It is also true, however, that the periodicity of the substrate structure plays a part in determining whether or not epitaxy occurs. The theory of heterogeneous nucleation does not at present take account of the nature of the substrate structure. It seems likely that the effects indicated by the theory may well play a part in the mechanism of epitaxial growth, but that other important factors may also supervene.

In experiments on the condensation of rare gases on amorphous carbon, Ball and Venables (1969) have obtained general confirmation of the theory of heterogeneous nucleation, as given by Walton and Rhodin and further developed by Lewis and Campbell (1967). Measurements of the saturation density of xenon nuclei as a function of temperature and rate of arrival of atoms, suggest that the critical nucleus contains only two atoms for temperatures below 38·5° K, and about 30 atoms for nuclei formed above this temperature. The dependence of the saturation density n_s on the number of atoms in the cluster i follows the relation

$$\frac{n_s}{n_0} = \left(\frac{R}{n_0 \nu}\right)^{i/i+1} \exp\left\{\frac{E_i + iE_d}{(i+1)kT}\right\} \qquad 4.8$$

(Joyce, Bradley and Booker, 1967), where E_d is the surface diffusion energy and E_i the binding energy for a cluster of i atoms.

Eqn. 4.8 indicates a linear relation between $\ln n_s$ vs $1/T$ with a slope which depends on i. Results of these experiments are shown in fig. 4.3, yielding the value of T at which the change from 2-atom to 30-atom nuclei occur.

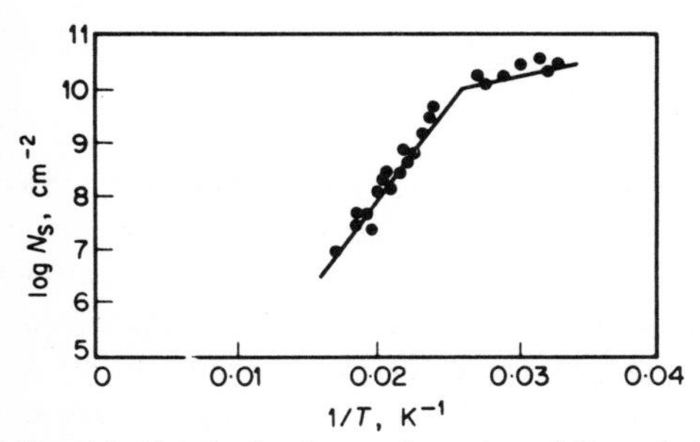

FIG. 4.3. Results for the condensation of films of xenon on carbon.

Among the many nucleation experiments which have been carried out on monocrystalline substrates, the results on silicon are of interest in illustrating both the general applicability of the theory of heterogeneous nucleation and also the way in which the periodicity of the substrate structure plays a part. We would intuitively expect the form of the distribution of the substrate surface atoms to influence the type of critical nuclei which form – an equilateral triangular nucleus sits in greater comfort on the (111) plane of a cubic material than on the (100) plane. On the latter plane, with its four-fold symmetry, nuclei of two or four atoms would be expected. These expectations are confirmed in studies of films on silicon. Although we do not have complete agreement between theory and experiment, the results are such that the general features of nucleation phenomena may be understood.

In the case of the rare gases on carbon, there is no evidence that preferred nucleation occurs at surface defects. In many experiments, however, and particularly in cases in which the binding energy of atoms to nuclei is large compared with the energy of adsorption to the surface, the results suggest that surface imperfections (e.g. cleavage steps, dislocation lines, F-centres, impurities) play an important role in determining nucleation phenomena. Early experiments on, e.g., alkali halide surfaces, tend to be difficult to interpret since the presence of adsorbed layers of residual gas, often present in uncertain amounts in such experiments, tends to mask the effects of true substrate forces. More recent work using surfaces of crystals which have been cleaved *in vacuo* indicates marked differences in behaviour compared with that on gas-contaminated surfaces.

4.3 *The effects of electron bombardment on film structure*

Exposure of the substrate surface to an electron beam during the deposition process is in some cases found to have a profound effect on the growth and structure of the resultant film. This constitutes one of many discoveries which create difficulties in interpretation of early work. Thus in experiments in which the source material is heated by electron bombardment, instead of by passing a current through a crucible, the substrate may well be subjected to bombardment by electrons, unless precautions are taken to avoid this. There are two dramatic effects produced by such bombardment. One is a marked increase in the density of nuclei on the surface and the second is a great improvement in the perfection of single-crystal films grown by epitaxy. Plate V shows the Kikuchi pattern obtained by Chambers and Prutton (1967) from an epitaxial film of nickel grown on vacuum-cleaved NaCl with a concomitant bombardment of 7 kV electrons. The profusion and sharpness of the Kikuchi lines indicate a very high degree of perfection in the film. It seems highly likely that the effect of the bombarding electrons is to produce a high density of surface defects which act as nucleation centres. It is significant that electron bombardment has no effect on substrates for which the energy required to create defects is very high.

4.4 *Post-nucleation growth*

Although in many instances, the general features of the structure of a film may be strongly influenced by the conditions of formation of the initial nuclei, it must not be inferred that subsequent growth processes are of little importance. There are some general rules inasmuch as oriented growth of the initial nuclei, which occurs under conditions for epitaxy, will generally lead to the formation, on subsequent growth, of a monocrystalline film. Growth on an amorphous substrate, with no preferred orientation of initial crystallites, will generally result in a polycrystalline film. There are, however, many post-nucleation processes which can occur and which can influence the final structure of the film. Mention has already been made of the evidence from cine-microscopy of a high mobility of *large* crystallites during the growth stages of a film. At an early stage, a large density of small crystallites forms. Later, the larger crystallites which have formed are separated by regions with a much lower density of crystallites than was present at the earlier stage. When the growing film is viewed in the electron microscope, small crystallites are seen to disappear extremely quickly and, simultaneously, a larger crystallite in the neighbourhood increases in size. Sometimes a whole large crystallite will, when it swallows a small one, change its orientation slightly. This is especially easy to see if the film is grown under conditions such that Moiré fringes are observed. The disappearance of the initial crystallites is too fast to follow but indicates that very rapid surface diffusion occurs.

At a later stage of growth, coalescence of crystallites is observed and in this process the particles often exhibit a liquid-like behaviour, even though the film temperature may be very far below the melting-point of the material. (Some care is needed in identifying the melting point of crystallites with that of the bulk. In general, the melting point of small crystallites is lower than that of the bulk.) Thus Plate VIII shows the structure observed by electron microscopy of a film of selenium in which two droplets are in the throes of coalescing. The striking thing about this specimen is that electron diffraction reveals it to consist of perfectly oriented single-crystals. The shapes of such coalescing crystallites may be satisfactorily accounted for by applying the simple ideas of droplet behaviour together with surface diffusion. Thus for the system shown in fig. 4.4 growing under the

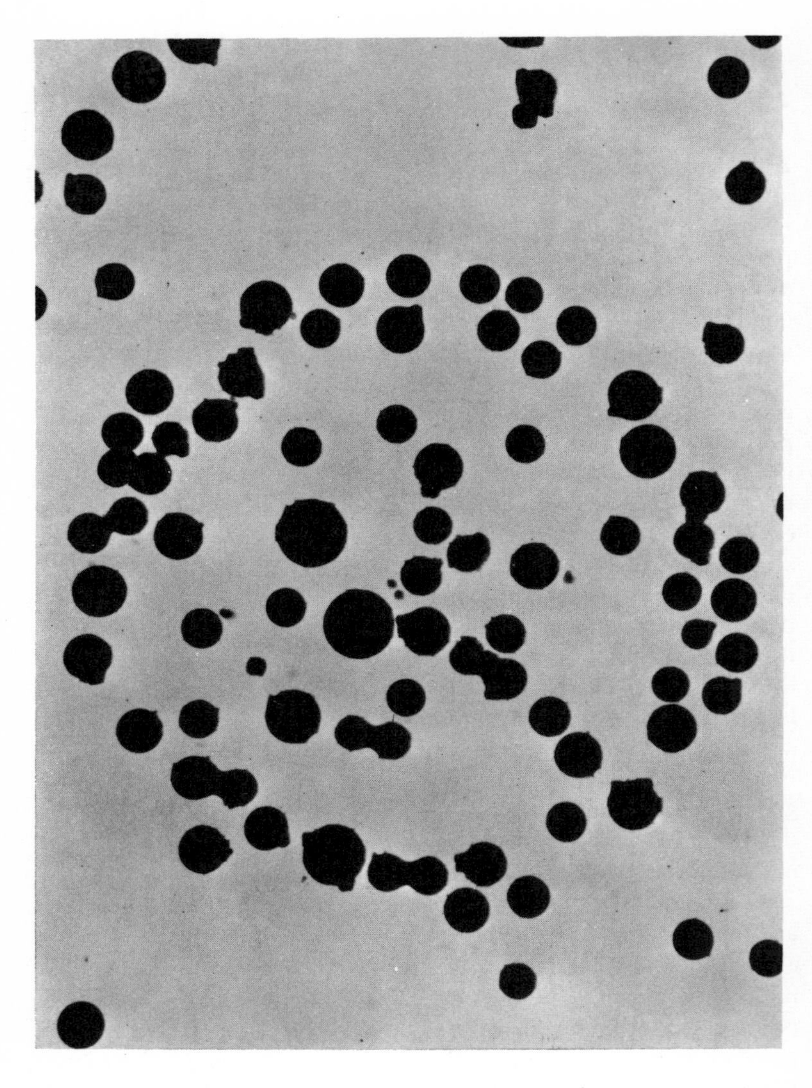

Plate VIII. Structure of selenium film, deposited on potassium bromide.

influence of atoms which arrive by surface diffusion, the relation between x, r and the time takes the form

$$\frac{x^6}{r^2} = \frac{25D\gamma\Omega^{4/3}t}{kT} \qquad 4.9$$

where D is the surface diffusion coefficient and γ the surface tension. The time to grow to a dumbell of given x/r is proportional to r^4, so that the coalescence of small droplets proceeds more rapidly than that of large ones.

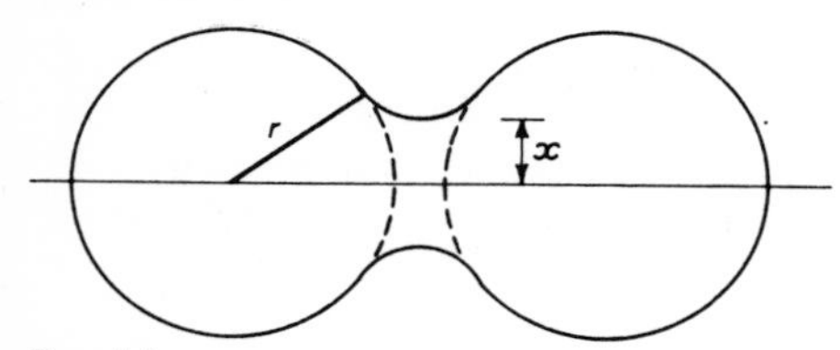

FIG. 4.4.

These ideas may further be extended to apply to the coalescence of crystallites with slightly different orientations. In this situation, a grain boundary will occur between the coalescing crystals in the region of the waist. There will be an interfacial energy associated with such a boundary and it is clear therefore that, if the shape of the system is such that movement of the boundary involves an increase in area, such a movement will not occur. As soon as the waist has filled out, as in fig. 4.5, the grain boundary is free to move

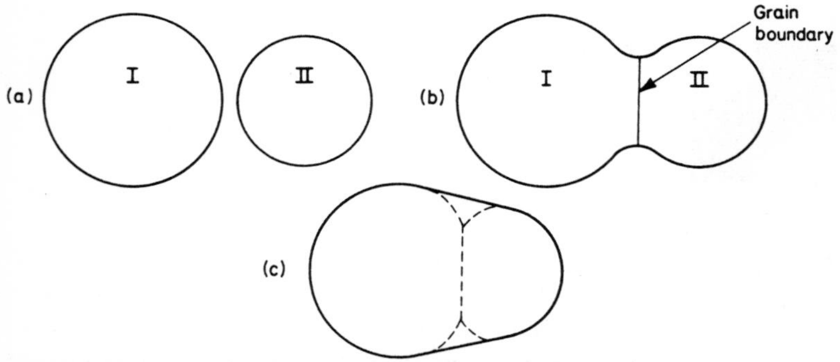

FIG. 4.5. When a grain boundary occurs as at (b), it cannot move laterally since its area would increase. Deposition occurs until stage (c) is reached, when the boundary moves.

and re-orientation of one part of the combined crystals to match that of the other can proceed. Precisely this behaviour is observed, through Moiré fringes, between such crystals and their substrate.

It has not so far proved possible to make cine-electron micrographs under ultra-high vacuum conditions. It is conceivable, therefore, that the mobile behaviour of microcrystallites referred to above is associated with gas contamination of the substrate surface. In the same way, the droplet shapes exhibited may arise from such contamination. There is some evidence that, under ultra-high vacuum conditions but for similar substrate temperatures and deposition rates, crystallites form with crystallographically orientated faces. In this situation (occurring, for example, in lead telluride films on rock-salt) the sequence by which crystallites coalesce is as shown in fig. 4.6. This behaviour can be accounted for on the assumption that the

FIG. 4.6. Schematic of coalescence in growing films of lead telluride.

growing crystallite takes up atoms from a catchment area determined by the range of surface diffusion. The shape of crystallite formed will depend on the relative binding energies for atoms on different faces and on the degree of (two-dimensional) supersaturation on different faces. Thus it is possible that the supersaturation on the (110) face of a cubic crystal will be larger than that on the (100) face. Starting with a multifacetted crystal such as that in fig. 4.7, the more rapid growth on the non-cube faces will lead to the development of crystallites bounded by (100) faces. The relation between the crystallite

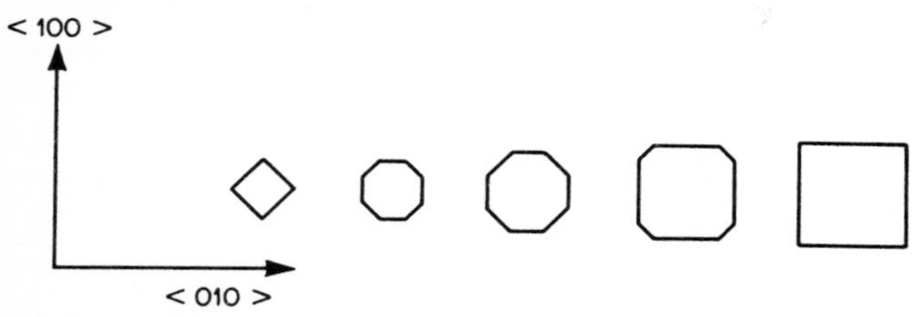

FIG. 4.7. Development of cube faces from initial dodecahedra.

size and thickness will depend on the relative rates of surface diffusion of the deposit atoms on the substrate surface and on the growing crystallite. When these are of the same order, one typically observes initially a mainly lateral growth of thin crystal plates up to a limiting size, followed by an increase in plate thickness and a slower rate of lateral growth. The crystallites finally merge together and may form a film of uniform thickness and with few dislocations. This is particularly likely if the adsorption energy of deposit molecules on the substrate is approximately equal to the binding energy of the deposit atom on the material itself.

4.5 *Film structures*

In view of the foregoing remarks and the variety of factors influencing film nucleation and growth, it is evident that only rather broad generalizations may be made of the types of structures existing in thin films. In terms of descending order of perfection, we may obtain the following:

(*i*) Films which are monocrystalline, continuous and with a very high degree of perfection, showing a low dislocation density. These are generally produced by deposition on a heated single-crystal substrate at an elevated temperature, possibly with accompanying irradiation by an electron beam.

(*ii*) Films which exhibit spot patterns by electron diffraction, consisting of a mosaic of orientated crystallites with relative misalignment of a degree or two.

(*iii*) Films which are polycrystalline but with a significant degree of alignment in which a preponderance of the crystallites are well orientated.

(*iv*) Polycrystalline films with no preferred orientation. This is frequently the structure observed in films deposited on non-crystalline substrates, or on crystalline substrates at low temperatures (e.g. room temperature or below).

(*v*) Polycrystalline films in which the crystallite size is so small that they exhibit very broad diffraction rings, in the limit showing patterns characteristic of an amorphous structure.

Progress in the study of the solid state made great strides when good specimens of single (bulk) crystals became available. In the same way our understanding of the behaviour of films has increased since the techniques of producing good single-crystal films have developed. Although the first examples of epitaxially grown films were produced in the 1920's, detailed studies of the physical properties of such films are of much more recent origin.

The polycrystalline film is a far more complicated system to analyse than a monocrystalline one, so that one cannot expect as complete an understanding of its properties as for a single-crystal film. However, polycrystalline films are of considerable technological importance and play a highly significant role in many fields. In subsequent chapters, the physical properties of films, both monocrystalline and polycrystalline, will be examined.

5

Mechanical Properties of Films

5.1 *General features*

The mechanical behaviour of films is of importance from two main points of view. In principle, an understanding of such behaviour can lead to a better understanding of the properties of bulk materials. In practice, the satisfactory operation of many thin film devices depends critically on the formation of stable films, which can withstand the rigours of a hostile environment.

As with many film properties, the mechanical properties of films depend on the usual multiplicity of factors attendant on their preparation. On account of experimental difficulties, the majority of the work done on mechanical properties has been carried out on polycrystalline films. Because of their greater structural complexity the results are more difficult to interpret than those for monocrystalline films. Some studies of epitaxial films have been made, but the nature of the rather delicate measurements entailed in extracting information on mechanical properties makes for difficulties in these studies.

Most of the studies made have been on metal films although attention has also been directed to dielectric materials, which are of importance in various electrical and optical devices. Measurements made include those of stress and strain, creep and plastic behaviour, fracture strength and, to a small extent, hardness. Various theoretical models have been proposed although at this stage detailed agreement between model and experiment is not in sight. There are, however, some general features which serve as a guide for future work.

When films are formed by thermal evaporation, or by vapour decomposition, on a heated substrate, then unless the film and substrate materials have identical expansion coefficients, a thermal stress will develop when the system cools to room temperature. This effect, which is present in many cases, may manifest itself dramatically in the form of detachment of the film from the surface. It is in fact not by any means clear that thermal stress effects will be absent in films deposited on room temperature substrates. The 'temperature' at which the film forms, inasmuch as this is not a somewhat ill-defined concept, may differ from the substrate temperature. Typically, the condensing atoms arrive with a very high thermal velocity: the effective 'temperature' of the condensing film depends on a balance of factors which govern the thermalization of the condensed material and are generally very difficult to assess. The temperature of the substrate surface will be partly determined by the radiation received from the source and partly by the latent heat given out by the condensing film. As the thickness of a metallic film increases, a large fraction of thermal energy incident on the substrate may be reflected. Moreover, since the optical constants of very thin films vary rapidly (and often in a complicated manner) with film thickness, this effect is very difficult to assess. Before discussing some of these effects in detail, we shall look at the experimental methods employed for studies of mechanical properties.

5.2 *Experimental techniques*

(*a*) STRESS AND STRAIN MEASUREMENTS

Measurements of the stress in films have generally been made by a beam-bending technique, in which the film is deposited on a thin rectangular beam and the deflection measured. Any of a variety of methods may be applied to the measurement of the small deflection which occurs – interferometric, capacitance and electro-mechanical arrangements have been employed. In most cases, the general solution for the bending of a composite beam, of two materials with different elastic properties, is not needed since the film thickness is usually small compared with the beam thickness. If the film is firmly bonded to the substrate and if no plastic flow occurs at the interface, then for a beam of thickness d, Young's modulus Y and Poisson's ratio σ, the stress S in a film of thickness t is given by

$$S = \frac{Yd^2}{6\rho t(1-\sigma)} \qquad 5.1$$

where ρ is the radius of curvature of the stressed beam, assumed initially straight.

The direct measurement of strain by a straight loading method has been applied in spite of the severe difficulties associated with the mounting of the films. A schematic of one of the systems used (designed originally for studies of whiskers, but adapted for film work) is shown in fig. 5.1. The solenoid/magnet combination imparts

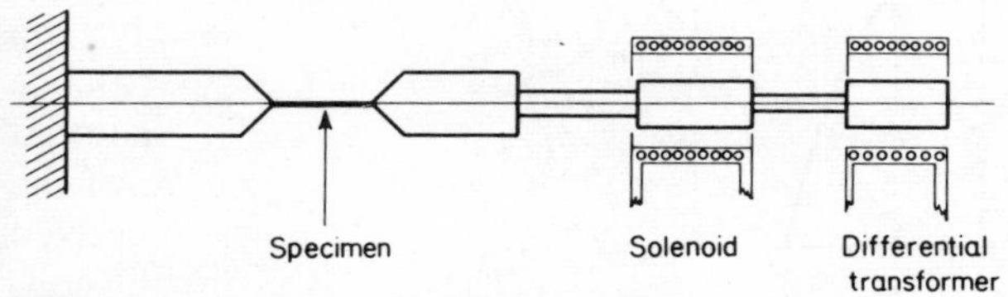

FIG. 5.1. Apparatus for stress/strain measurements.

a known stress to the film and the extension of the film is recorded by the differential transformer. In other cases optical methods have been used to measure the small extensions observed. Generally methods of this type are of use only for fairly thick films – say above about the 0·1 μm region – on account of the severe handling problems for thinner specimens. An alternative to the 'linear loading' method is that in which the film is mounted on the end of a cylinder and is made to bulge by applying a differential pressure. Interference fringes formed between the film and a reference flat provide a sensitive method of measuring the displacement. To a crude approximation, it may be assumed that the film deforms into a spherical cap. If a is the radius of the film, T_0 the tension in the film with zero pressure difference and ρ the radius of curvature for an applied pressure p then

$$p = \frac{2t}{\rho}\left[T_0 + \frac{Ya^2}{3\rho^2(1-\sigma)}\right] \qquad 5.2$$

and

$$T = T_0 + \frac{Ya^2}{3\rho^2(1-\sigma)} \qquad 5.3$$

where t is the film thickness, Y is Young's modulus and σ is Poisson's ratio for the film material. In fact the form of the film surface approximates more nearly to a quadratic surface of revolution and a better measure of the radius of the film at the apex is obtained from two measurements (fig. 5.2) of the height of the film above the

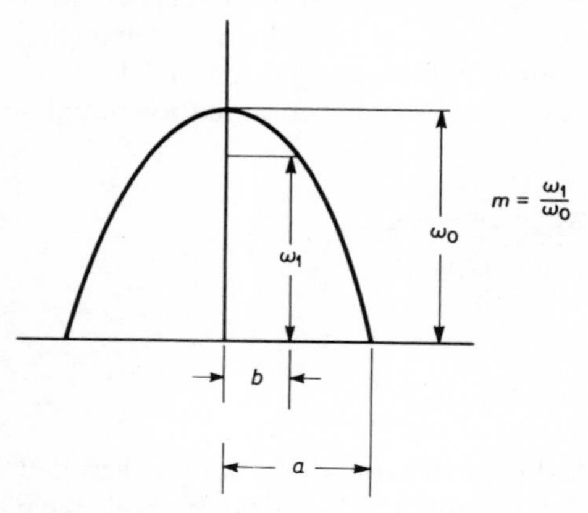

FIG. 5.2.

base, one at the apex and the other at the quarter chord. With the notation of fig. 5.2 the radius of curvature is given by

$$\rho = \frac{a^2}{2w_0}\left[\left(\frac{b}{a}\right)^2 - (1-m)^2\right] \Big/ m(1-m). \qquad 5.4$$

The bulge method of studying films has been applied in a neat way to epitaxially grown films of gold. In this case, the film can be grown over the entire surface of a rocksalt crystal and a hole made in the substrate by a high-pressure water jet. The remainder of the film stays anchored to the crystal and the usual problems of attaching the film are avoided.

Although in principle strain in the microcrystallites of a film may be determined by the broadening of electron or X-ray diffraction rings or spots, it is often difficult to separate with certainty such effects from those of crystallite size and from instrumental artefacts.

(*b*) BREAKING STRAIN MEASUREMENTS

Two methods have been employed to study the fracture of films under stress. In the first of these, applied to the study of electro-deposited films, a cylindrical rotor is spun at a sufficiently high speed to cause fracture. (It is ensured that the adhesion between film and rotor is low). In the second method, used inside the electron microscope, the film is fixed to a slotted plate (fig. 5.3) the ends of

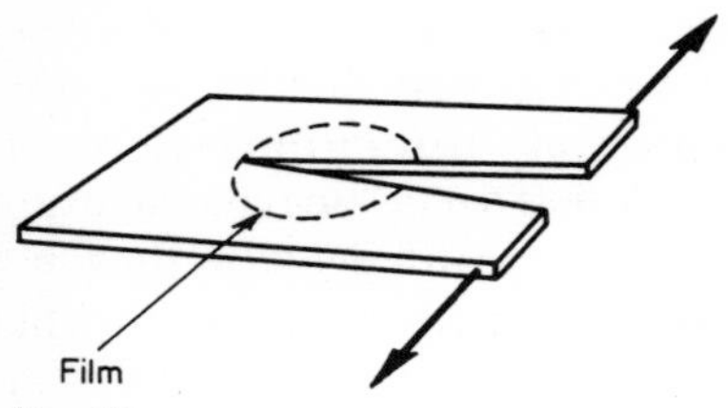

FIG. 5.3.

which are forced apart. Although only approximate breaking stresses can be obtained in this way, this method has the advantage that the film structure and the movement of dislocations may be observed during the experiment.

5.3 *Stress in films formed by thermal evaporation* (Hoffman, 1966)

As indicated above, a state of stress is expected when a film is deposited on a heated substrate if the thermal expansion coefficients of the film and substrate materials differ. However, it is found that

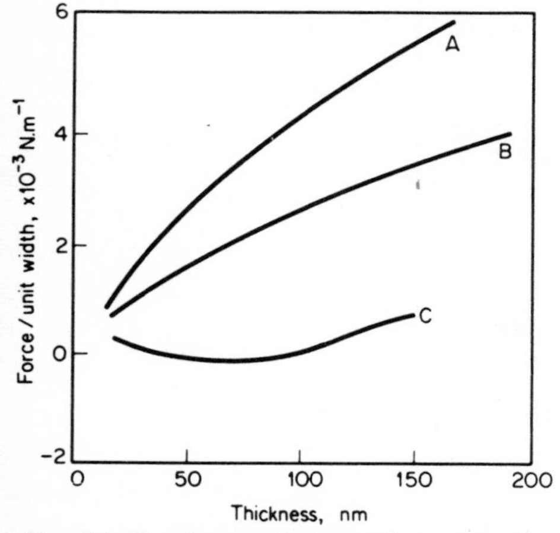

FIG. 5.4. Tensile stress in thermally evaporated iron films: A: mean substrate temperature 100° C. B: 165° C. C: 240° C.

over and above the stress resulting from differential contraction, an intrinsic stress occurs in many materials. Typical results for iron films are shown in fig. 5.4, in which the stress arising from differential contraction effects has been subtracted from the total observed stress. The magnitude of the intrinsic stress depends both on film thickness and on substrate temperature and probably arises from features of the film structure. Since films are often found to curl on removal from the substrate, it is clear that the stress varies with depth in the film, again probably reflecting a variation of film structure with depth. Both compressive and tensile intrinsic stresses are found in different materials, often with a highly complicated dependence on substrate temperature. Thus Permalloy W films exhibit a tensile stress for deposition below about 100° C and above about 350° C and a compressive stress between these temperatures. The actual temperature at which the change-over from tensile to compressive stress occurs depends to some extent on the rate of formation of the film. There is at present no detailed theory to account for this general behaviour.

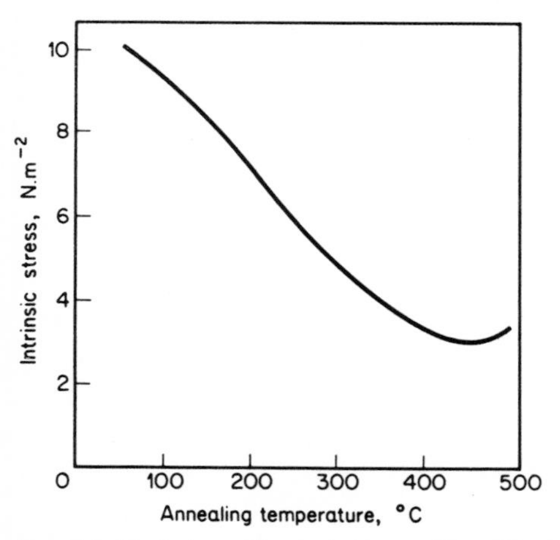

FIG. 5.5. Variation of intrinsic stress of iron films with annealing temperature.

The effect of annealing is generally to reduce the intrinsic stress although, as indicated in the curves of fig. 5.5, there may be an optimum temperature for minimum stress.

5.4 *Elastic and plastic behaviour of films*

Studies of the stress-strain behaviour of films often yield a low value of elastic modulus for the initial loading followed by normal values on subsequent unloading and reloading. It is not entirely certain whether the initial behaviour is associated with creep or slip in the methods used to hold the film. Results obtained by the bulge method show a similar behaviour, although the modulus for the initial loading is nearer to the bulk value than that for the tensile machine method. Polycrystalline films formed by thermal evaporation generally yield elastic moduli close to that of the bulk material; on the other hand, low moduli have been observed for chemically deposited films and also for electrolytic films. In the case of chemically formed films the difference in behaviour is probably due to impurities in the film.

So far as creep behaviour is concerned, this is still an open question. There is evidence both from epitaxial gold films and from chemically-deposited films that creep does not occur. Completely elastic behaviour is exhibited up to the point at which the film breaks. There are two reasons why the observation of apparent creep may occur. One is due to an initial straightening of the film and the second to slipping of the film in the holders. Although it is possible that some observations may be accounted for in this way, there seems nevertheless to be evidence that genuine creep does occur in evaporated films in much the way in which it is known to occur for rolled foils and other 'bulk' specimens.

At sufficiently high stresses, localized plastic deformation of a film occurs, leading to a reduction in film thickness and a consequent rise in stress level. Dislocations nucleate at intercrystallite boundaries, move along glide planes and eventually produce microscopic cracking of the film. The stress level at which this occurs is in many cases far higher than that typically observed in annealed bulk specimens and often significantly higher than that for drawn or cold-worked material. The breaking stress has been measured by Beams (1959) in an ingenious way, in which the film is deposited on a cylindrical rotor which is then spun at an increasing rate until the film breaks. The bulge technique mentioned above has also been employed. It is found that for polycrystalline silver and nickel there is generally a dependence of breaking strain on thickness. No such thickness-dependence is observed for copper. For gold films, there is conflicting evidence: some observers find a thickness-dependence whilst others do not.

We can see in a general way why the strength of films may be higher than that of bulk materials. The failure of a specimen depends on the propagation of dislocations and the limited thickness of a film severely restrains such motion. However, no detailed theoretical model exists which can account for the wide variety of results obtained. In some cases films have been prepared in rather poor vacua, so that there may be effects due to oxide on the film surface. In the same way that the strength of certain metal whiskers is believed to be associated with surface oxide, so may the high strength of films be accounted for. That there are other factors involved is clear from the case of gold, which also shows a high strength, but which should be free from surface oxide.

When films are deposited on amorphous substrates at normal incidence, the stress is found to be isotropic. Anisotropic stress distribution is observed in films formed by deposition at non-normal incidence. It is known from electron microscopic studies that asymmetry exists in the structure of such films, so the occurrence of stress anisotropy is not surprising.

In view of the remarkable low-friction properties of polytetrafluoroethylene (PTFE) and of the fact that friction is essentially a surface phenomenon, it is of interest to enquire whether films of this material can be used as reducers of friction. It is found that coatings of PTFE can be made by sputtering, that they adhere very well to materials such as mild or silver steel and that the coefficient of friction, μ, is comparable to that for the bulk material. As would be expected, the observed value of μ rises if the thickness of the coating is of the order of or less than the scale of the surface roughness of the specimen coated.

The films so produced differ somewhat from the bulk material. They are amorphous and are significantly harder than the (crystalline) bulk material. For a given load, the value of μ rises very slightly with increasing thickness (fig. 5.6), once the thickness for complete coverage is reached. The value of μ recorded, for a given thickness of coating, decreases with increasing load, as shown in fig. 5.7. The load quoted is that applied to a coated cylinder, 12·5 mm diameter by 62·5 mm long, lying on a flat plate, the friction of the line contact being measured.

On heating films of PTFE to 450° C for 15 minutes, a slight decrease is observed in the value of μ, due possibly to the development of crystallinity in the film.

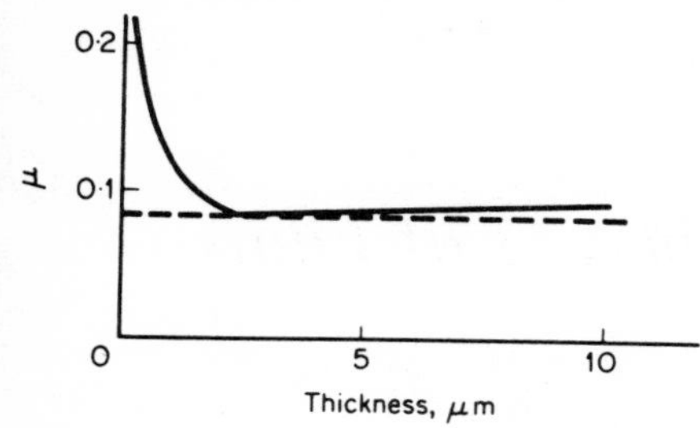

FIG. 5.6.

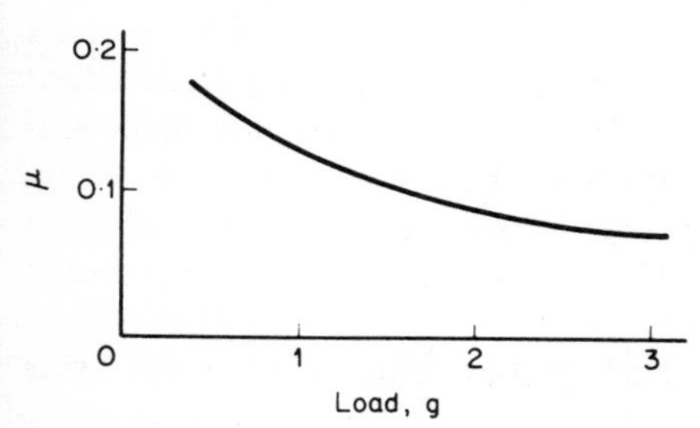

FIG. 5.7.

6

Optical Properties of Films

6.1 *General comments*

Probably the earliest scientific application of thin film properties was that involving the use of silver films for mirrors and for instruments such as the Fabry-Perot interferometer. The methods used for the production of such films now appear to have a culinary slant and to have involved the skill of the cooking expert for successful results. Nevertheless the importance of the early work, particularly in the realm of spectroscopy, was paramount; the accuracy of measurements on spectral lines far surpassed that of other physical measurements.

Interest in the silver films used in these early experiments was limited simply to their possessing suitable optical properties – high reflection and low absorption – for interferometry. In this chapter we shall discuss the results of extensive studies of recent times in which the detailed optical behaviour of films, both metallic and dielectric, are examined in relation to their thickness, structure and conditions of preparation. The story is very far from being complete so far as concordance of numerical values of film parameters is concerned. In respect of some properties, there are almost as many different values of parameters as there have been measurements made. We are, however, beginning to see in a general way how optical behaviour of films – particularly in the case of very thin metallic films – depends on film structure. The vagaries of the properties of dielectric films have been less severe than those of metals although even here we do not always have a clear picture.

The optical measurements generally made on films include reflectance and transmittance, either at normal or non-normal incidence and the state of polarization of the reflected light may be measured over a range of wavelengths. More rarely, determinations of scattering loss in films have been made.

The reflective and transmissive properties of films are readily calculated from electromagnetic theory, in terms of the refractive index and thickness of the film material. (As will be seen below, this blithe assumption that a real film may be so simply characterized is not always valid.) In the case of absorbing films some cumbrousness is involved. In particular, although reflectances and the state of polarization of a reflected beam may be calculated explicitly in terms of the optical parameters of the film, the reverse is not the case.

6.2 *Reflectance and transmittance of single film*

For the general case of light incident at an angle ϕ_0 on a film of index n_1 and thickness d_1 (fig. 6.1), lying on a plane substrate of

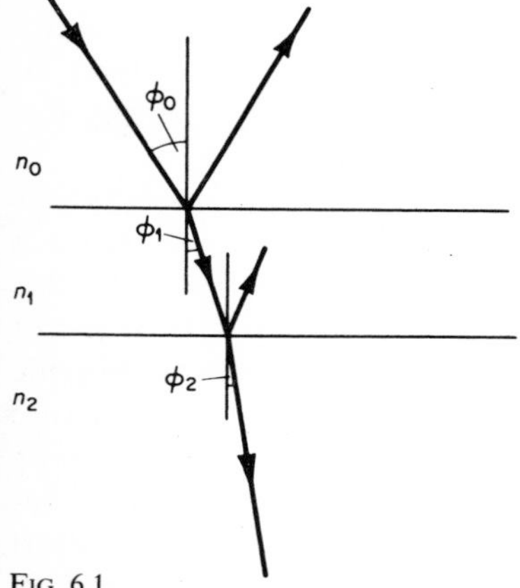

FIG. 6.1.

index n_2, the reflectance and transmittance may be conveniently determined in terms of the Fresnel coefficient of reflection and transmission at the n_0/n_1 and n_1/n_2 interfaces. It is most convenient to work with components of the light beams resolved in and perpendicular to the plane of incidence. Thus if the electric vector of the incident beam lies in the plane of incidence, the Fresnel (amplitude reflectance and transmittance) coefficients are given by:

$$r_{1p} = \frac{n_1 \cos\phi_0 - n_0 \cos\phi_1}{n_1 \cos\phi_0 + n_0 \cos\phi_1} \qquad t_{1p} = \frac{2n_1 \cos\phi_1}{n_1 \cos\phi_0 + n_0 \cos\phi_1}. \qquad 6.1$$

For light polarized with the electric vector perpendicular to the plane of incidence, the corresponding quantities are:

$$r_{1s} = \frac{n_1 \cos\phi_1 - n_0 \cos\phi_0}{n_1 \cos\phi_1 + n_0 \cos\phi_0} \qquad t_{1s} = \frac{2n_1 \cos\phi_1}{n_1 \cos\phi_1 + n_0 \cos\phi_0}. \qquad 6.2$$

The reflectance and transmittance are easily determined by summation of the multiply reflected and transmitted beams, taking account of the phase difference, $\delta_1 = 2\pi\nu n_1 d_1 \cos\phi_1$, between successive beams. ν is the wavenumber. It should be noted that in general n_0, n_1 and n_2 may all be functions of ν. In the case of a film in air the variation of n_0 with ν may be neglected. Since we are characterizing the film by a single parameter n_1, this result will apply only to an isotropic film. For an anisotropic film it is necessary to take into account the dependence of refractive indices on direction of the beams.

The result of the summation, for an isotropic film, is given by:

$$\mathbf{R} = \frac{r_1^2 + 2r_1 r_2 \cos 2\delta_1 + r_2^2}{1 + 2r_1 r_2 \cos 2\delta_1 + r_1^2 r_2^2} \qquad \mathbf{T} = \frac{n_2}{n_0} \frac{t_1^2 t_2^2}{1 + 2r_1 r_2 \cos 2\delta_1 + r_1^2 r_2^2} \qquad 6.3$$

where the Fresnel coefficients for either s- or p-components are inserted.

The variation of **R** with δ_1 for films of various indices, on a substrate of index $n_2 = 1{\cdot}5$ and in air ($n_0 = 1$), is shown in fig. 6.2 for the case of normal incidence. This may be regarded as the variation of **R** with thickness for a given wavenumber ν. *If* dispersion is ignored, it may also be regarded as the variation of **R** with wavenumber for a given thickness. The effects of dispersion for many real materials over the visible spectrum are not large.

If films are sufficiently thick for two or more maxima to be obtained in a spectrophotometric curve for normal incidence, then the optical thickness $n_1 t_1$ (neglecting dispersion) may be simply obtained. If successive maxima occur at ν_N, ν_{N-1} then since $2n_1 t_1 = N/\nu_N = (N-1)/\nu_{N-1}$, both N and $n_1 t_1$ may be found. The reflectance at the maxima ($\mathbf{R}_{max}$) or minima ($\mathbf{R}_{min}$), are given by

$$\mathbf{R}_{max} = \left(\frac{n_2 - n_0}{n_2 + n_0}\right)^2 \qquad \mathbf{R}_{min} = \left(\frac{n_1^2 - n_0 n_2}{n_1^2 + n_0 n_2}\right)^2 \qquad 6.4$$

for the case $n_0 \lessgtr n_1 \lessgtr n_2$ and

$$\mathbf{R}_{max} = \left(\frac{n_1^2 - n_0 n_2}{n_1^2 + n_0 n_2}\right)^2 \qquad \mathbf{R}_{min} = \left(\frac{n_2 - n_0}{n_2 + n_0}\right)^2 \qquad 6.5$$

for $n_0 \gtrless n_1 \lessgtr n_2$. Thus in principle a determination of the reflectance at the turning values may yield the values of n_1, and even some

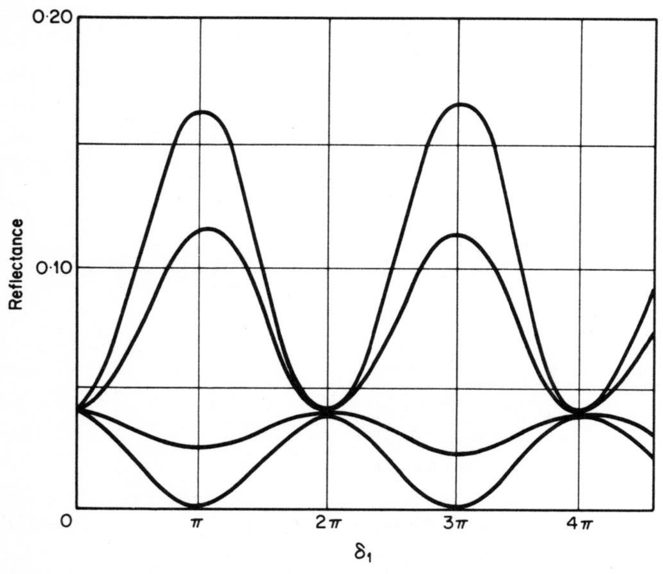

FIG. 6.2. Reflectance vs phase thickness. ($\mathbf{R}$ vs δ_1 for given ν or – if dispersion is neglected – $\mathbf{R}$ vs ν for given d_1.)

estimate of the dispersion may be made. This method should, however, be used with circumspection since various factors may, in a real film, conspire to cause errors. In particular, scattering loss, which increases rapidly as shorter wavelengths are approached, may give low maxima and incorrect derived indices.

In some cases, the results obtained experimentally agree reasonably well with those predicted by equations 6.4 and 6.5. If the film consists of a homogeneous, parallel-sided slab of bulk material on a substrate of lower index, then the index n_1 calculated from $\mathbf{R}_m$ will agree with the bulk figure and the reflectances at the maxima will correspond to that of the uncoated substrate. This behaviour is illustrated in fig. 6.3 by magnesium fluoride on glass. In contrast,

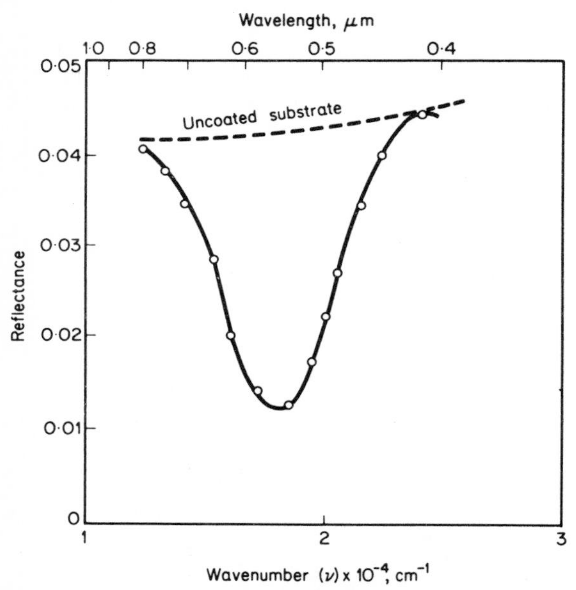

FIG. 6.3. Reflectance of magnesium fluoride film.

however, we may note the behaviour of calcium fluoride, evaporated under moderate vacuum conditions, which yields the curve of fig. 6.4. In this case the observed behaviour may be accounted for on the assumption of a gradient of refractive index normal to the plane of the film.

6.3 *Absorbing films*

If the material of the film is absorbing, then some modification of the above expressions is needed. The description for the case of

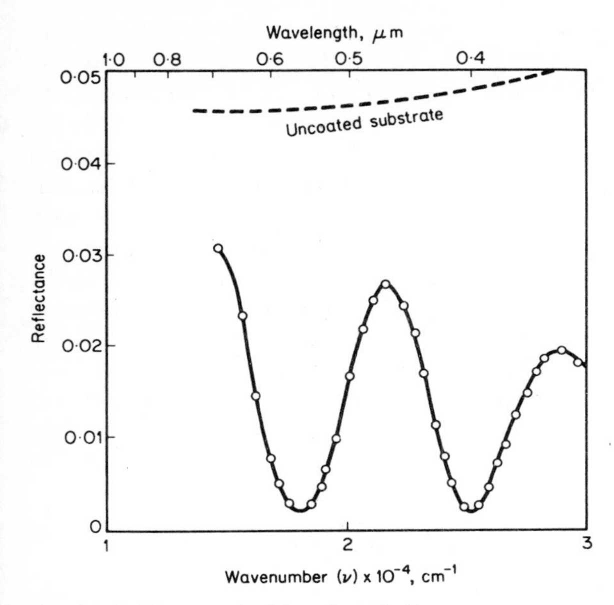

FIG. 6.4. Reflectance of calcium fluoride film.

non-normal incidence becomes highly involved, particularly in respect of the wave actually propagating inside the film. In a transparent medium, the form of the propagation is such that planes of constant phase lie parallel to planes of constant amplitude. For an isotropic medium, these are perpendicular to the direction of the ray. In an absorbing medium, light entering at non-normal incidence propagates with planes of constant amplitude inclined to those of constant phase – described as an *inhomogeneous* wave. The mathematical description is cumbersome and of little interest so far as films are concerned. It is generally sufficient to study absorbing films *either* by transmission at normal incidence *or* in reflection. In the former case the description is simple, since for this special case, planes of constant phase are parallel to those of constant amplitude. In the latter case, the reflected radiation, which is homogeneous, is examined.

For a wave entering an absorbing medium at normal incidence, the propagating wave may be described by a complex index $(n_1 - ik_1)$. The form of the wave may therefore be written for propagation in the z-direction, as

$$\exp 2\pi i\left[\nu t-(n_1-ik_1)\frac{z}{\lambda_v}\right] \equiv \exp\left(-\frac{2\pi k_1}{\lambda_v}\right)\exp 2\pi i\nu\left(t-\frac{n_1 z}{c}\right)$$

when λ_v is the vacuum wavelength of the radiation. The physical significance of k_1 is that for a length of path equal to one vacuum wavelength, the attenuation is $\exp(-2\pi k_1)$.

Expressions for the reflectance and transmittance of an absorbing film may be obtained by substituting (n_1-ik_1) in place of n_1 in the expressions for the Fresnel coefficients (eqns 6.1 and 6.2), performing the summation and evaluating the square of the modulus of the resulting amplitudes. The results are given in Section 6.7 below. The difference between spectrophotometric curves for transparent and absorbing films is illustrated by the case of Sb_2S_3, for which results are given in fig. 6.5. On the high frequency side of the absorption

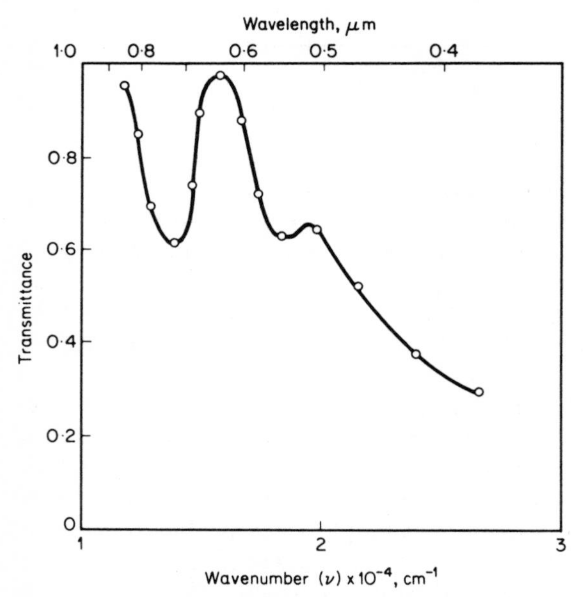

FIG. 6.5. Transmittance of antimony sulphide film.

edge, the film is non-absorbing and the curve is similar to that of fig. 6.2. In the region of the edge, the film becomes absorbing and the transmittance curve becomes damped.

It will be clear that if the absorption coefficient of the material of a film is sufficiently high, the effects of multiple reflections in the film will be negligible and the transmission curve will fall monotonically with increasing thickness. This is in fact the behaviour exhibited by metal films in the visible region: the value of k_1 is so high that the transmittance of films 100 nm thick is generally less than 1%. (In the far ultraviolet region, however, at frequencies exceeding the plasma frequency of the electrons, metal films may exhibit a near-transparent behaviour.)

6.4 *Optical constants of film materials*

(*a*) TRANSPARENT FILMS

Refractive indices of film materials may be determined from spectrophotometric curves, as indicated above. Alternatively, the simple method introduced by Abelès (1950) may be used. If light polarized with the electric vector parallel to the plane of incidence falls on a partly-covered surface at an angle of incidence $\phi_B = \text{arc tan}\, n_f$, then the reflectance of the filmed and uncoated parts of the surface are equal. An extra lens in the telescope of an ordinary spectrometer enables the line of demarcation at the edge of the film to be clearly seen and an intensity match to be made. The reflectance-equality condition is independent of the refractive index of the substrate. The condition is also independent of film thickness, although for certain thicknesses, the discrimination is poor (fig. 6.6).

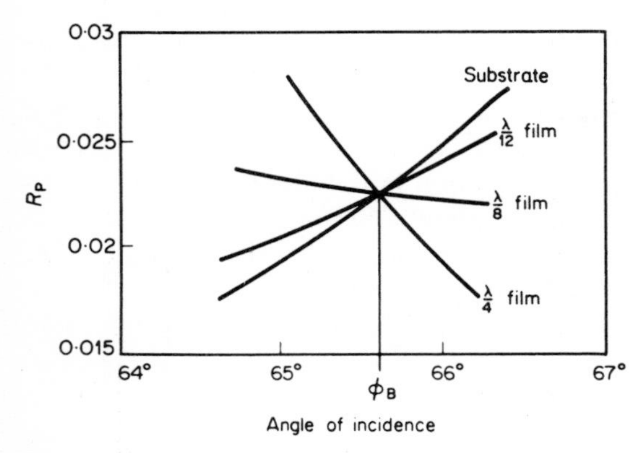

FIG. 6.6.

Measurements of refractive indices of materials in film form indicate a variety of structures. In some cases, compact films form, with refractive indices equal to those of the bulk materials. Magnesium fluoride is a good example of such a material, even showing a slightly higher figure than that reported for mineral MgF_2 (Sellaite), possibly due to some CaF_2 content in the latter. In other cases, the refractive indices recorded are significantly lower than those of the bulk, due to the formation of an aggregated structure containing voids. Refractive index measurements on such films, both dry and with immersion techniques, enable the extent of voids to be determined. The porosity of films such as lithium fluoride is often found to depend on film thickness, due to the growth, in thick films, of large orientated crystallites.

Careful measurements of the reflecting and transmitting properties of films which form with refractive index less than the bulk value generally reveals that they cannot be represented as macroscopically uniform layers. Although the evolution of a highly detailed model is generally somewhat difficult, the general characteristics, e.g. of refractive index gradients or the presence of low-index transition layers, can generally be deduced. These are in fact frequently artefacts associated with the conditions of deposition. Thus the refractive index n_f of CaF_2 films may be made low enough to satisfy the condition $n_f = \sqrt{n_g}$ (where n_g is the index of glass) which ensures complete antireflection for, say, the green region of the spectrum. This is achieved by evaporation in poor vacuum; the films so formed are of little practical value on account of their fragility.

Transparent dielectric films play an important part in providing multilayer optical systems, with a range of important properties for the optical designer. Some of the more useful systems are described below. The range of wavelengths over which they may be used is restricted by the onset of absorption in the film material. This sets the present ultraviolet limit of about 250 nm for such systems. In the infrared region, the limit is generally set by the fact that stable layers of sufficiently large thickness cannot be made.

Table 6.1 gives details of a selection of some of the more important dielectric films in common use.

(*b*) ABSORBING FILMS

The determination of the (two) optical constants of absorbing iso-

TABLE 6.1

Material	*Range of transparency*		*Refractive index*
	from	*to*	
Cryolite	<200 nm	10 μm	1·35–1·39
MgF_2	230 nm	10 μm	1·39–1·40
SiO_2	200 nm	5 μm	1·43–1·45
SiO	500 nm	8 μm	1·55–1·95 (depending on rate of evaporation and on O_2 and H_2O pressure)
ZnS	400 nm	14 μm	2·37 at 500 nm
CeO_2			2·42 at 550 nm
Ge	1·7 μm	20 μm	4·3–4·6
PbTe	3·8 μm	—	5·5

tropic films is somewhat more difficult than that of determining the refractive index of a transparent film. For a detailed survey of methods, the author's article in the *Physics of Thin Films*, **2** (1964), may be consulted. One or two methods will be briefly outlined here.

If several films of different thicknesses are made and if the thicknesses are determined independently (e.g. by the use of multiple-beam Fizeau fringes) then use may be made of the fact that for thicknesses large enough for multiple reflections to be ignored, the transmittance of the film is given by:

$$\mathbf{T} = \frac{16n_0 n_2(n_1^2+k_1^2)^2}{[(n_1+n_2)^2+k_1^2][(n_0+n_1)^2+k_1^2]} \exp(-4\pi\nu k_1 d_1). \qquad 6.6$$

Thus a plot of $\ln T$ vs thickness yields a straight line (beyond a certain thickness) from the slope of which the value of k can be found. A measurement of the reflection coefficient for a thick ($T \simeq 0$) film enables n to be found from the measured reflectance and k-value since, for normal incidence,

$$\mathbf{R} = \frac{(n_1-n_0)^2+k_1^2}{(n_1+n_0)^2+k_1^2}. \qquad 6.7$$

If the film to be examined lies on a transparent substrate, so that the reflectance $\mathbf{R}'$ on the substrate side may be measured, then this together with the transmittance $\mathbf{T}$ and the reflectance $\mathbf{R}$ on the air side enables the optical constants and thickness to be determined. The detailed procedure is described by Malé (1950). (It should be noted that whereas the transmittance $\mathbf{T}$ is the same regardless of the direction of the incident beam, the reflectance depends on the refractive index of the adjacent medium.)

One of the most sensitive methods of making optical measurements is that of using polarimetry. It is seen from equations 6.1–6.3 that the amplitude reflection coefficient for a beam incident at a given angle on a surface (or a film) depends on the state of polarization. Considering the incident beam resolved into p- and s-components, the values of R_p and R_s differ. Likewise the phase change on reflection (the argument of R) also depends on the state of polarization so that in general an incident plane polarized beam is reflected in a state of elliptical polarization. Analysis of the ellipticity of the reflected beam can lead to an evaluation of the optical constants of the surface or film. Whether the procedure involved is merely tedious or hideous depends on the nature of the surface or film. Thus if we first consider a surface we find that life is tolerable if the optical constants are such that $n^2+k^2 \gg 1$. This condition is fulfilled by most metals in the visible region. (It is not fulfilled by semi-conductors in the near infrared.) To within the above approximation, the optical constants are simply related to the angle of principal incidence Θ (the angle of incidence for which the reflection ellipse has major and minor axes parallel and perpendicular to the plane of incidence) and the angle of principal azimuth Ψ (angle between plane of incidence and direction of resultant when quarter-wave plate is inserted in reflected beam). Under this condition, n and k are given by

$$\begin{aligned} n &= \tan\Theta \sin\Theta \cos 2\Psi \\ k &= \tan 2\Psi \end{aligned} \qquad 6.8$$

Such a method may be applied to a very thick film. For a thin film the state of polarization of the reflected light may be determined from equation 6.22 (below) but unhappily the equations (for the s- and p-components) cannot be inverted to give n and k in terms of the measured Θ and Ψ. Moreover, the computation of the values of Θ and Ψ for a given set of values of n, k and thickness is somewhat tedious. (This arises because the $\cos\phi$ terms in the equations become complex.) With the arrival of the electronic computer, this problem has been somewhat simplified, since a programme may be written and sets of curves, e.g. of Θ vs Ψ with one curve for each thickness, may be produced. Alternatively, a compensator may be used so that the differential phase change Δ on reflection at the surface is measured and thus, with the restored azimuth ψ then constitute the two measured parameters. The thickness is measured independently. Fig. 6.7 shows a typical curve, due to Archer (1962)

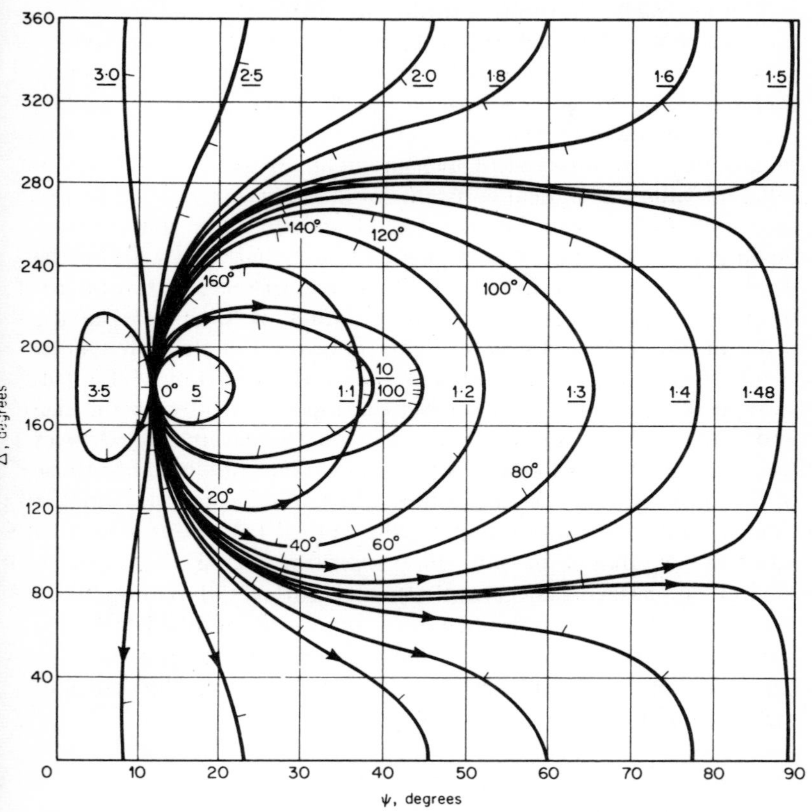

FIG. 6.7. Δ, ψ for transparent films on silicon. Figure underlined is film index. Phase thickness δ in degrees marked on each curve. Film thickness is $1{\cdot}517\delta/(n^2 - 0{\cdot}883)^{1/2}$ nm.

for films on silicon. With a certain amount of good fortune, the curves for different thicknesses do not cross, so that a given set of measurements of ψ, Δ and thickness enable n and k to be determined unequivocally.

Early measurements of the optical constants of metals ran into the difficulty that the method of preparing the specimen surface largely determined the results obtained. This is so because the value of k for metals is such that the penetration of an incident wave is confined to the first hundred or so nm. Since it was impossible to devise a method which would give a surface which was smooth to this order, such variations were hardly surprising. The rise of the method of vacuum evaporation appeared to provide a way out of this difficulty. Glass surfaces could be produced with a sufficiently high degree of smoothness to serve as substrates and films could be deposited in vacuum, leading surely to smooth, thin layers of pure metal, ideal for this type of measurement. In fact, the number of different values of optical constants obtained was roughly equal to the number of different investigators. The assumption that one was dealing with a homogeneous, plane-parallel slab was wide of the mark. The films were found, on the arrival of the electron microscope, to be aggregates of microcrystallites of 'diameters' of the order of nm or tens of nm. Straight application of electromagnetic theory to such a system (which had been done in 1904 by McLaurin) showed that the optical behaviour depended critically on the size and shape of the crystallites. This is so even if we assume the particles to consist of chunks of metal with optical constants equal to those of the bulk. In fact, for crystallites whose dimensions are smaller than the electron mean-free-path, the optical behaviour would be different from that of the bulk, quite apart from shape effects. Owing to electron scattering effects at the particle surfaces, the conductivity and hence the extinction coefficient would be modified from the bulk values. In this situation, the lack of concordance between experimental values and theoretical predictions from the theory of metals should not be taken too seriously.

6.5 *Inhomogeneous films*

Interest is awakening in films which are characterized by a variation of refractive index in the direction normal to the plane of the film. Such films can be made by simultaneous evaporation from two sources, at differently controlled rates. Thus if the rate of deposition of component A (index n_A) varies linearly with time from 0 to R_A whilst that of B (index n_B) varies from R_B to 0 during the same interval, the refractive index will vary with position (z) in the film according to the relation:

$$n^2(z) = \frac{n_A^2 - \alpha n_B^2}{(1-\alpha)} + \frac{\alpha(n_B^2 - n_A^2)}{(1-\alpha)[1+(\alpha^2-1)z/d]^{1/2}} \qquad 6.9$$

where $\alpha \equiv R_B/R_A$ and d is the film thickness.

By the use of rate-monitoring devices consisting of quartz oscillators whose resonant frequency varies with the mass of the deposited film, the rates of deposition may be varied according to any prescribed curve and a variety of refractive index profiles obtained.

The optical behaviour of a general homogeneous film could be estimated by treating it as a large number of sub-layers and applying the theory of multilayers given in section 6 below. For an arbitrary profile, this is about the only easy method. For certain specific profiles, however, the wave equation can be solved explicitly. Thus for the case where the ratio $R_A/R_B = 1$, the expressions for the electric (E) and magnetic (H) fields in the layer may be written

$$\begin{aligned} E &= A\xi^{1/2}J_{1/3}(2kd\xi^{3/2}/a) + B\xi^{1/2}J_{-1/3}(2kd\xi^{3/2}/a) \\ H &= -\frac{1}{iZ_0}[A\xi J_{-2/3}(2kd\xi^{3/2}/a) - B\xi J_{2/3}(2kd\xi^{3/2}/a)] \end{aligned} \qquad 6.10$$

where $\xi = az/d + n_A^2$, $a = n_0^2 - n_A^2$, $Z_0 = \sqrt{(\mu_0/\varepsilon_0)}$ and A, B are known constants involving the rates R_A, R_B (Jacobsson, 1963).

The reflectance of a single inhomogeneous film, formed under the above conditions, is a function both of the ratio R_A/R_B and of the refractive index of the substrate. Fig. 6.8 shows curves of constant reflectance for the parameters R_A/R_B and substrate index n_s. For this diagram, the refractive indices of the constituents are taken as $n_A = 1{\cdot}35$, $n_B = 2{\cdot}35$. A feature of particular interest is that there is a considerable area of the R_A/R_B vs n_s diagram for which the value of reflectance is below 0·5%. This indicates that, for a reasonable range of substrate indices and evaporation rate ratio, low reflectance is obtainable. This is of interest, particularly since it indicates that very precise control of the evaporation rates is not essential.

6.6 *Multilayer systems*

(*a*) THEORY

During the twenty years after the Second World War, considerable

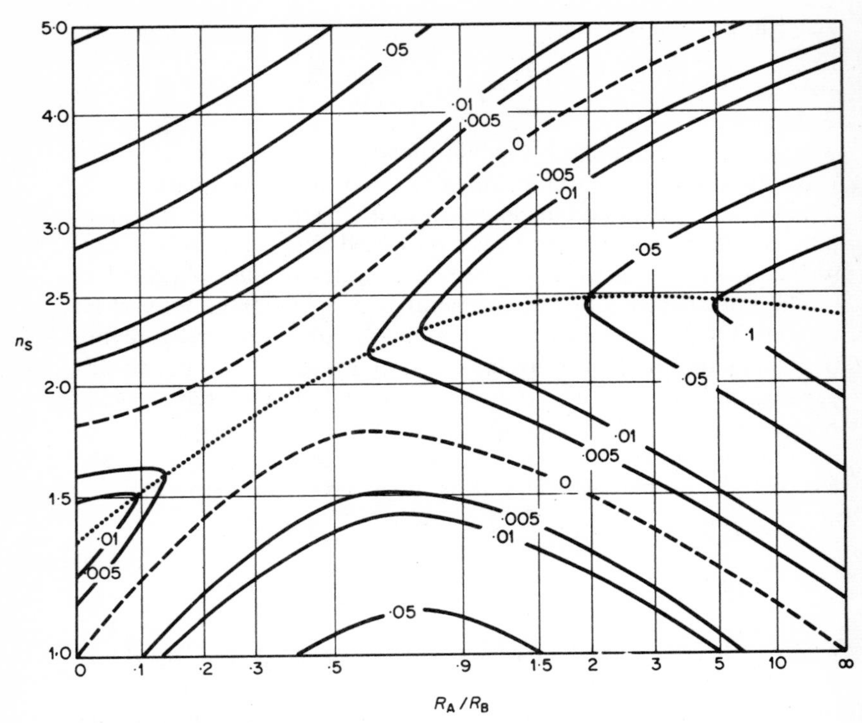

FIG. 6.8. Curves of constant reflectance for various R_A/R_B and n_s.

development of multilayer systems of films took place, in a manner which has had a considerable impact on many areas of optics. Most attention has been confined to systems of plane-parallel-sided, trans-transparent, homogeneous layers with thicknesses generally of the order of the wavelength of the radiation used. (Evaporated films for the optical region; slabs of plastic for the microwave region!)

Derivation of the expressions for the reflectance and transmittance of a stratified system is straightforward. The usual electromagnetic boundary conditions are applied at each interface so that the amplitude of the waves (both directions) in any layer may be readily written in terms of those of the adjacent layers. Thus by successive operations, the amplitude of the waves emerging from either side of the system may be expressed in terms of that of the incident wave.

The final results are readily expressed in terms of the product of a set of 2×2 matrices. There are two possible methods of handling the problem. In one case, each matrix is a function of the Fresnel coefficients at an interface and of the thickness of the adjacent film. In the other case each matrix is a function of the film thickness and refractive index (or effective indices, p- and s-, for the case of non-normal incidence. These are $n_\nu/\cos\phi_\nu$ for the electric vector in the plane of incidence and $n_\nu \cos\phi_\nu$ for $E_\nu \perp$ plane of incidence).

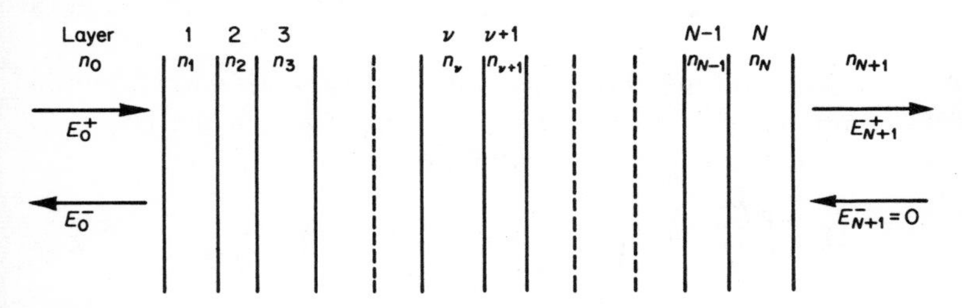

FIG. 6.9.

In the first-named case, we have, for the system of fig. 6.9,

$$\begin{pmatrix} E_0^+ \\ \overline{E_0^-} \end{pmatrix} = \frac{\prod_{\nu=1}^{N+1} M_\nu}{\prod_{\nu=1}^{N+1} t_\nu} \begin{pmatrix} E_{N+1}^+ \\ \overline{E_{N+1}^-} \end{pmatrix} \qquad 6.11$$

where t_ν is the Fresnel transmission coefficient for the interface between the νth and $(\nu+1)$th layer and the matrix M_ν is given by

$$M_\nu = \begin{pmatrix} \exp i\delta_{\nu-1} & r_\nu \exp i\delta_{\nu-1} \\ r_\nu \exp(-i\delta_{\nu-1}) & \exp(-i\delta_{\nu-1}) \end{pmatrix}, \qquad 6.12$$

δ_ν is the phase thickness of the νth layer and r_ν the Fresnel reflection coefficient at the $\nu/\nu+1$ boundary.

For the light incidence from the n_0 side, we have $E_{N+1}^- = 0$ and the amplitude reflectance E_0^-/E_0^+ and transmittance E_N^+/E_0^+ are easily found to be

$$R = \frac{m_{21}}{m_{11}} \qquad T = \frac{\prod_{\nu=1}^{N+1} t_\nu}{m_{11}} \qquad 6.13$$

where m_{11}, m_{21} are the elements of the product matrix, i.e.

$$\prod_{\nu=1}^{N+1} M_\nu = \begin{pmatrix} m_{11} & m_{12} \\ m_{21} & m_{22} \end{pmatrix}. \qquad 6.14$$

The energy reflection and transmission coefficients are then given by

$$\mathbf{R} = RR^* \qquad \mathbf{T} = \frac{n_{N+1}}{n_0} TT^* \qquad 6.15$$

The alternative method of derivation is to work with the total field components $E_\nu = E_\nu^+ + E_\nu^-$, $H_\nu = H_\nu^+ + H_\nu^-$ in the layers, in which case the recurrence relation becomes

$$\begin{pmatrix} E_{\nu+1} \\ H_{\nu+1} \end{pmatrix} = \begin{pmatrix} \cos\delta_\nu & (i\sin\delta_\nu)/n_\nu \\ in_\nu \sin\delta_\nu & \cos\delta_\nu \end{pmatrix} \begin{pmatrix} E_\nu \\ H_\nu \end{pmatrix}. \qquad 6.16$$

From boundary conditions at the n_0/n_1 interface, we have $E_{\nu+1} = E_{\nu+1}^+$ and $H_{\nu+1} = n_{\nu+1} E_{\nu+1}^+$. Successive application of equation 6.16 thus yields the necessary relationship between E_{N+1}^+ and E_0^+, E_0^-, so that the reflectance and transmittance may be found.

In either case, hand computation becomes tedious for more than a very small number of layers. The evaluation of 2×2 matrix products is, however, a trivial computer operation so that multilayer computations will generally be handled in this way.

(*b*) TYPICAL MULTILAYER SYSTEMS

One of the most generally used systems of multilayers is that of the high-reflecting stack, consisting of a stack of layers of equal optical thickness and alternating refractive index. For the wavelength λ_0 for which the layer thicknesses are $\lambda_0/4$ (or an odd multiple of this), all the multiply-reflected beams are in phase. The system has a high reflectance at λ_0 and has, for a large number of layers, a flat top. The width of the reflectance band (fig. 6.10) is given by

$$\frac{\Delta\nu_0}{\nu_0} = \frac{4}{\pi}\arcsin\frac{n_H - n_L}{n_H + n_L} \qquad 6.17$$

and the reflectance at ν_0 is given by

$$\mathbf{R}_{2N} = \left(\frac{n_s f - n_0}{n_s f + n_0}\right)^2 \text{ for an even number}$$

and $$\mathbf{R}_{2N+1} = \left(\frac{fn_{\mathrm{H}}^2 - n_0 n_{\mathrm{s}}}{fn_{\mathrm{H}}^2 + n_0 n_{\mathrm{s}}} \right)^2 \text{ for an odd number}$$

of layers, where n_0 is the refractive index of the incident medium, n_{s} that of the substrate and $f = (n_{\mathrm{H}}/n_{\mathrm{L}})^{2N}$.

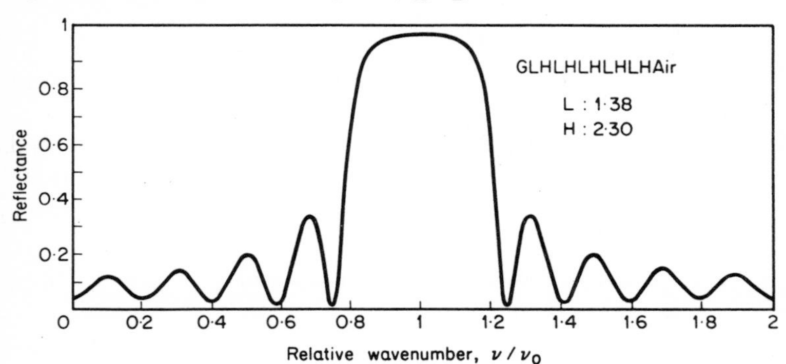

FIG. 6.10. Reflectance band of quarter-wave stack.

It is readily seen that, for moderate numbers of layers of typical dielectric film materials, very high reflectances should in principle be attainable. Thus taking $n_{\mathrm{H}} = 2{\cdot}42$ and $n_{\mathrm{L}} = 1{\cdot}38$ (values for CeO_2 and MgF_2 respectively for the green) we find that for a 15-layer stack, $\mathbf{R} = 0{\cdot}9996$. In practice, reflectances tend not to attain the theoretical values since some scattering loss occurs at inhomogeneities in the deposited layers. Nevertheless, reflectances comfortably in excess of 99% are regularly produced in these systems.

The availability of high-efficiency reflections such as those discussed above leads immediately to the feasibility of making narrow-band transmission filters of the Fabry-Perot type in which a half-wave layer is sandwiched between high-reflecting stacks. Such filters were hitherto made with silver as the reflector. However, the high absorption loss in silver layers compared with all-dielectric stacks leads to a low-peak transmission of the filter.

The transmittance of a system consisting of a spacer layer enclosed by two reflecting layers whose reflectance is independent of wavelengths is the well-known Airy curve. Since the reflectance of a dielectric stack falls to low values at the edges of the reflectance band a somewhat different transmission curve is obtained, as shown

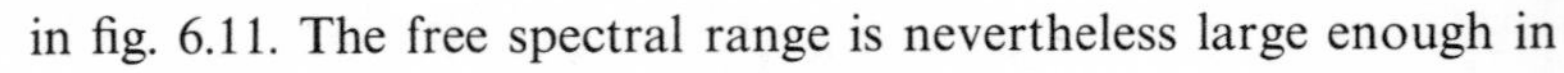
in fig. 6.11. The free spectral range is nevertheless large enough in

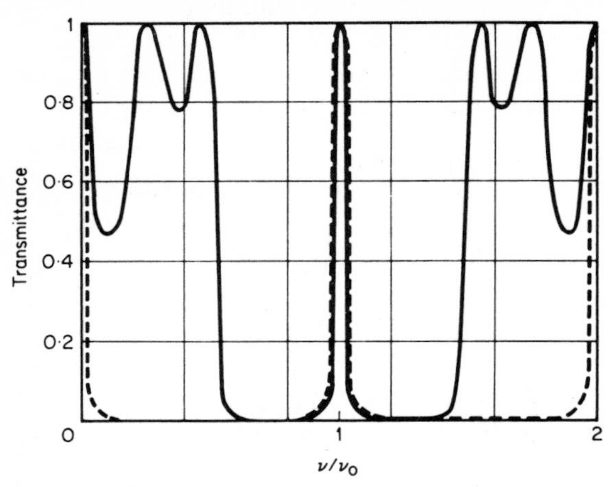

FIG. 6.11. Transmittance of all-dielectric Fabry-Perot type interference filter. The dotted curve would apply if the reflectors had constant reflectance.

practice for such filters to be of considerable value. Bandwidths down to less than one nm are obtainable, with peak transmission above 80% or so and excellent rejection outside the pass band.

In some respects, the Fabry-Perot system is not the best for narrow-band filter performance. The use of multiple half-wave layers such as those in figs. 6.12 and 6.13 lead to more square band tops. Moreover, the precise form of the transmission curve is less susceptible to distortion through errors in layer thickness than is the Fabry-Perot type.

One other important class of all-dielectric systems is that for reducing the reflectance of a dielectric/air interface. A single layer of optical thickness $\lambda_0/4$ and of refractive index $\sqrt{n_s}$ will reduce to zero the reflectance at a surface of index n_s at the wavelength λ_0. However, for wavelengths in the neighbourhood of $\lambda_0/2$, the reflectance of the surface is practically the same as that of the substrate. Mention has already been made of the possibilities existing with inhomogeneous films for antireflecting purposes. An alternative is to use a suitable stack of dielectric films, designed to give a series of zeroes and a low average reflection over an extended wavelength range. Thus fig. 6.14 shows what can be achieved with a system of four layers. The reflectance over the whole of the visible spectrum is held below ~0·5%.

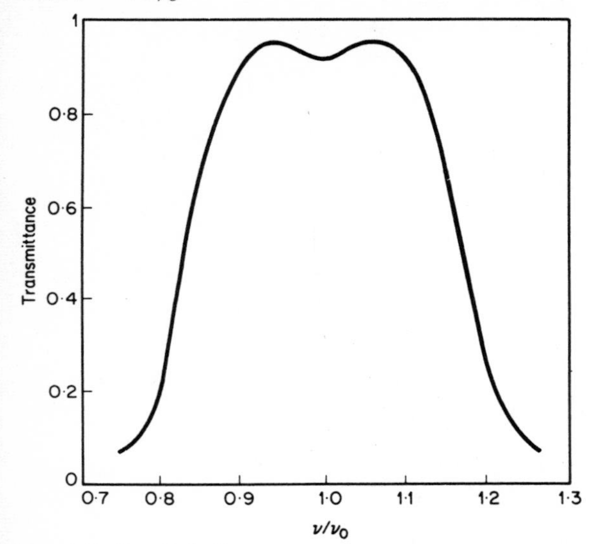

FIG. 6.12. Double half-wave system.

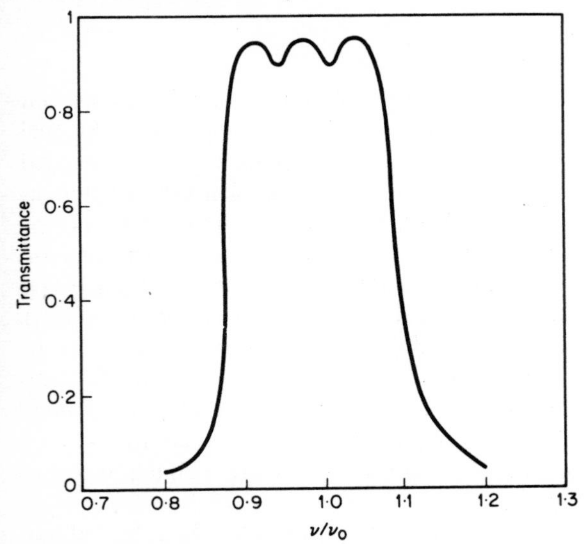

FIG. 6.13. Treble half-wave system.

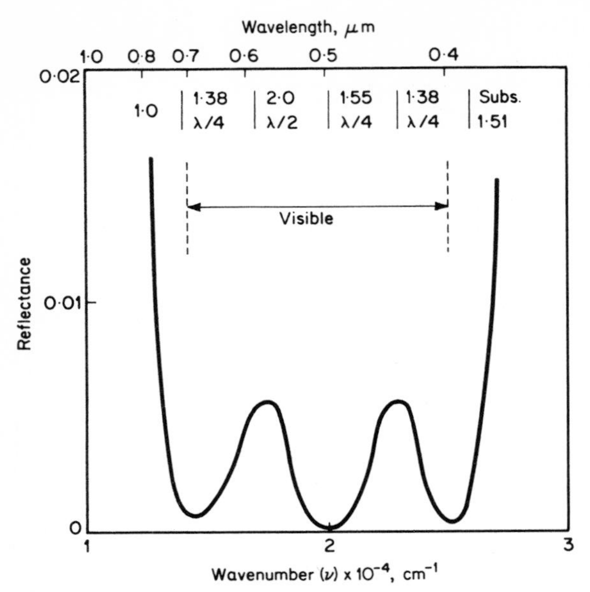

FIG. 6.14. Four-layer antireflecting system.

6.7 *Absorbing layers*

(*a*) SINGLE LAYERS

Formally, the discussion of absorbing materials may be carried out by substituting the complex quantity $n_j - ik_j$ in place of the real refractive index n_j used in the foregoing sections. Caution is needed here inasmuch as the propagation index is variously written as $n+ik$, $n(1 \pm ik)$, $\nu(1 \pm ik)$, etc. With the index in the form $n-ik$, the extinction coefficient k represents the effect of absorption; on traversing a distance in the medium equal to one vacuum wavelength, the amplitude of the wave decreases by the factor $\exp(-2\pi k)$.

In general, we are interested in the radiation reflected by or transmitted by the absorbing film and thus are free of the complexities that arise in discussing the form of the electromagnetic disturbance *inside* the absorbing material, where except for the case of normal incidence, planes of constant phase are inclined to planes of constant amplitude.

Even for the computation of the amplitudes for the reflected and transmitted waves, however, the calculation becomes fairly cumbersome, the effect of the innocent substitution of $n-ik$ for n giving rise to some awkwardness, especially for non-normal incidence. This arises in part from the fact that when Snell's law is applied, a complex value emerges for the 'angle' of refraction – a manifestation of the fact that the wave within the film is inhomogeneous.

When the film is thick enough for multiple reflections to be negligible (which is generally true for metal films more than a few tens of nm thick), the reflectances $\mathbf{R}_p$ and $\mathbf{R}_s$ may be written as:

$$\mathbf{R}_p = \frac{\mu + p\cos^2\phi_1 - (\mu + \sin^2\phi_1)\{2(\mu+\lambda)\}^{1/2}\cos\phi_1}{\mu + p\cos^2\phi_1 + (\mu + \sin^2\phi_1)\{2(\mu+\lambda)\}^{1/2}\cos\phi_1} \qquad 6.18$$

$$\mathbf{R}_s = \frac{\mu + \cos^2\phi_1 - \{2(\mu+\lambda)\}^{1/2}\cos\phi_1}{\mu + \cos^2\phi_1 + \{2(\mu+\lambda)\}^{1/2}\cos\phi_1} \qquad 6.19$$

where the medium of incidence has unit refractive index and

$$p = n^2 + k^2; \quad q = n^2 - k^2; \quad \lambda = q - \sin^2\phi_1; \quad \mu^2 = \lambda^2 + 4n^2k^2;$$

ϕ_1 is the angle of incidence of the beam.

For the case of a film of index $n_1 - ik_1$ on a substrate of index n_2, and where multiple reflections cannot be ignored, the expressions given in Section 6.2 may be applied. For this case, the Fresnel coefficients are complex and are written, for the case of *normal incidence*:

$$r_1 \equiv g_1 + ih_1 = \frac{n_0^2 - n_1^2 - k_1^2}{(n_0+n_1)^2 + k_1^2} + i\frac{2n_0k_1}{(n_0+n_1)^2 + k_1^2} \qquad 6.20$$

$$r_2 \equiv g_2 + ih_2 = \frac{n_1^2 - n_2^2 + k_1^2}{(n_1+n_2)^2 + k_1^2} - i\frac{2n_2k_1}{(n_1+n_2)^2 + k_1^2}. \qquad 6.21$$

Writing $2\pi\nu k_1 d_1 \equiv \alpha_1$ and $2\pi\nu n_1 d_1 = \gamma_1$, we find

$$\mathbf{R}_1 = \frac{(g_1^2+h_1^2)e^{2\alpha_1} + (g_2^2+h_2^2)e^{-2\alpha_1} + A\cos 2\gamma_1 + B\sin 2\gamma_1}{e^{2\alpha_1} + (g_1^2+h_1^2)(g_2^2+h_2^2)e^{-2\alpha_1} + \cos 2\gamma_1 + D\sin 2\gamma_1} \qquad 6.22$$

where $\quad A = 2(g_1g_2 + h_1h_2); \qquad B = 2(g_1h_2 - g_2h_1);$

$\quad C = 2(g_1g_2 - h_1h_2); \qquad D = 2(g_1h_2 + g_2h_1).$

An important point to notice in connection with absorbing films is that the value of the reflectance depends on the medium of incidence and that the reflectances at the two sides of an absorbing film are generally different. The general form of the reflectance and transmittance versus thickness for metallic films is typified by the curves of fig. 6.15, which are calculated for the optical constants of

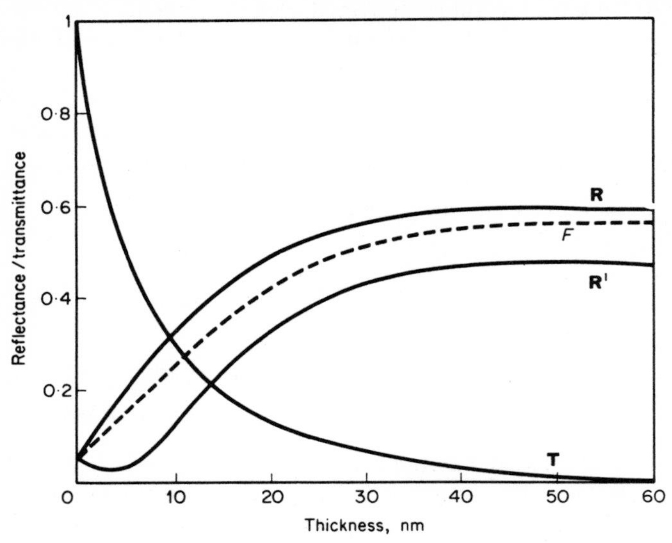

FIG. 6.15. Platinum. Full lines: results calculated from bulk optical constants. Dotted line: reflectance (air side) of sputtered platinum films.

platinum for $\lambda = 546$ nm. The experimentally observed results for platinum films show minor differences from the curves calculated from the bulk constants although this behaviour is exceptional, particularly in the case of metal films with thicknesses below 10–20 nm. The structure revealed by the electron microscope indicates why bulk behaviour is not observed. The films consist, in the extremes of very small thickness, of a dispersion of isolated metal particles. Thus the solving of Maxwell's Equations for a pair of adjoining continua is of singular irrelevance to the case of the real film.

Long before the optical behaviour of metal films became of interest, Maxwell Garnett calculated the behaviour to an electromagnetic wave of a random distribution of spherical conducting particles embedded in a dielectric medium, for the case where the particle diameter is small compared with the wavelength – a reasonable approximation to certain types of metal film. The results may be specified in terms of the volume fraction q occupied by the conducting particles and indicate that the system possesses an effective index $n_e - ik_e$, related to the bulk index $n - ik$ of the particles by the relation

$$\frac{(n_e - ik_e)^2 - 1}{(n_e - ik_e)^2 + 1} = q\frac{(n - ik)^2 - 1}{(n - ik)^2 + 1}. \quad 6.23$$

Since the derivation assumes that the usual Lorentz-Lorenz polarization correction may be applied – implying that the film thickness is very large compared with the particle diameter, it is not to be expected that the theory will apply to very thin layers, in which a two-dimensional distribution of particles is a better model. Moreover, the difficulty also arises as to what values of n and k to use for metal particles whose diameter is of the order of the electron mean-free-path. Certainly the bulk value will *not* be applicable. For thicker films, the general ideas of the Maxwell Garnett theory appear to be correct, although agreement in detail is lacking, probably due to the unwarranted simplicity of the spherical particle model. For use with very thin films, a development of the Maxwell Garnett theory has led to reasonable agreement. In this treatment, by Schopper, the model used is that of a two-dimensional distribution of oblate ellipsoidal particles. Measurements on gold films, of thicknesses in the range 1–6 nm show good agreement with the Schopper theory.

In view of the increasing evidence that the precise structure of metal films in small thicknesses depends critically on the conditions of preparation, it would appear that the search for models to represent this or that particular film is an exercise of perhaps limited usefulness. Sufficient evidence has, however, accumulated to show that the methods discussed above can be expected to yield reasonable results.

(*b*) ABSORBING LAYERS IN MULTILAYER SYSTEMS

When an absorbing layer forms part of a multilayer stack, the amplitude of the standing wave field in the layer will depend on the admittances of the stacks surrounding the absorbing layer. Now the amount of energy absorbed by the metal film depends directly on the amplitude of the field present. One has, therefore, a measure of control over the energy absorbed and can make it large or vanishingly small, for a given wavelength. The potentialities of this process appear considerable: one should be able to reflect selectively and with high efficiency at one wavelength and transmit freely at another.

Combinations of metal and dielectric stacks have not so far found extensive use except in a few special fields.

6.8 *Interpretation of optical absorption in alkali halide films*

Measurement of the optical properties of solids over a wide frequency range can yield considerable information on the electronic band structure of the material and some progress has been achieved in studies of the alkali halides. The spectral region from which much information is to be had is that covering the ultraviolet, down to wavelengths of the order 100 μm. Since the absorption coefficient in most of this region is very high, it is necessary to use thin films to obtain absorption spectra in transmission. Moreover, in order to avoid the effects of thermal broadening of absorption lines, measurements must be made at low temperatures. Certain features are displayed by absorption spectra obtained in this way which can be interpreted in terms of excitons. Thus the general form of the spectra of RbI, RbBr and RbCl (Teegarden, 1968) are shown in fig. 6.16. The significant feature is that of strong doublets, at the following energies:

RbI: 5·74, 6·95 eV
RbBr: 6·64, 7·12 eV
RbCl: 7·54, 7·66 eV

Now the ions of the halogens are isoelectronic with the rare gas atoms, whose spectra are well understood. In particular, it is known that absorption doublets occur in this region as a result of spin-orbit coupling and that the spin-orbit energies for the halogen ions concerned are: I^- : 0·95 eV, Br^- : 0·50 eV and Cl^- : 0·10 eV. These figures agree approximately with the peak separations given above. This would correspond to a hole in a p-orbital on the negative ion, to which a bound electron in an s-state is attached, thus giving rise to a p^5s configuration.

Some of the other features of the spectra of these and other alkali halide films can be interpreted in this way. There is, however, a difficulty which arises in connection with reflectance measurements which have been made on such films. From these measurements, together with the application of Kramers-Kronig relations, it is possible to deduce the absorption spectrum. Unfortunately, absorption spectra obtained in this way are often totally different from those obtained directly. The reasons for these discrepancies are not yet known.

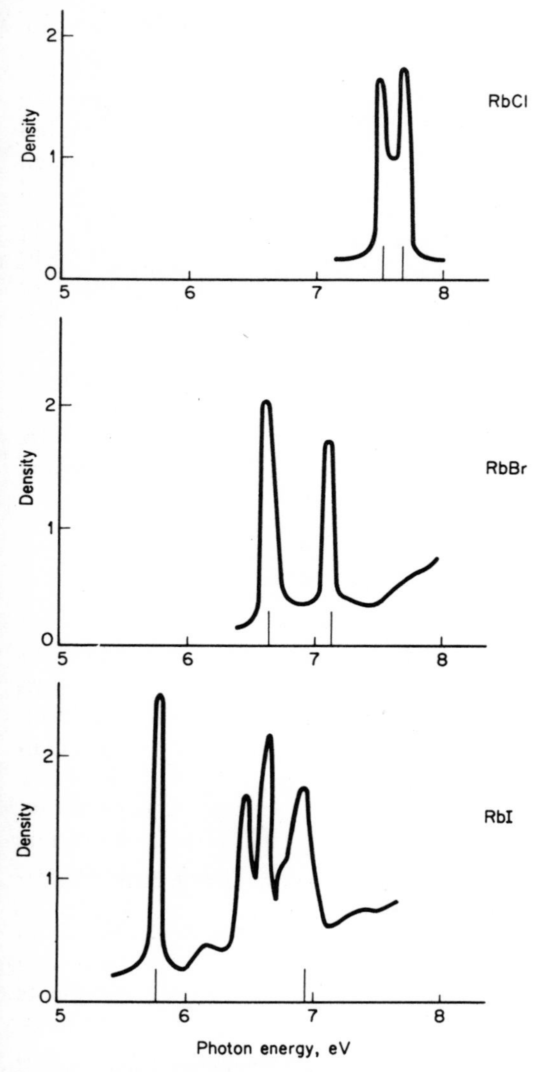

FIG. 6.16. Doublet structure in ultraviolet absorption spectra of alkali halide films.

6.9 *Anisotropic and gyrotropic films*

It is of interest (and of increasing importance) to examine the behaviour of anisotropic films (uniaxial or biaxial) and also of gyrotropic films. The commonest example of a gyrotropic medium is one which exhibits Faraday rotation of plane-polarized light. (In fact, such a medium transforms plane-polarized light into elliptically polarized light with the major axis inclined to the original direction of polarization. The ellipticity is, however, very small.)

Replacement of the isotropic films of the foregoing discussion by anisotropic *non*-gyrotropic films lead to some increase in algebraic complexity but no more. The most general case of a stack of absorbing, biaxial layers with optic axes in arbitrary directions would be lengthy to deal with (and of possibly little interest when completed!), but involves little more than determining, for each layer and for the appropriate ray direction, the appropriate values of the refractive index and extinction coefficient. To take the simple example of a uniaxial, non-absorbing film lying with its optic axis perpendicular to the substrate surface, with indices n_a and n_b we find for the Fresnel coefficients:

$$\begin{aligned}
r_{1s} &= \frac{n_0 \cos\phi_0 - (n_a^2 - n_0^2 \sin^2\phi_0)^{1/2}}{n_0 \cos\phi_0 + (n_a^2 - n_0^2 \sin^2\phi_0)^{1/2}} \\
r_{1p} &= \frac{n_a n_b \cos\phi_0 - n_0(n_b^2 - n_0^2 \sin^2\phi_0)^{1/2}}{n_a n_b \cos\phi_0 + n_0(n_b^2 - n_0^2 \sin^2\phi_0)^{1/2}} \\
r_{2s} &= \frac{(n_a^2 - n_0^2 \sin^2\phi_0)^{1/2} - n_2 \cos\phi_2}{(n_a^2 - n_0^2 \sin^2\phi_0)^{1/2} + n_2 \cos\phi_2} \\
r_{2p} &= \frac{n_2(n_b^2 - n_0^2 \sin^2\phi_0)^{1/2} - n_a \cos\phi_2}{n_2(n_b^2 - n_0^2 \sin^2\phi_0)^{1/2} + n_a \cos\phi_2}
\end{aligned} \qquad 6.24$$

from which, with the help of the expressions in Section 6.2 the values of **R** and **T** may be found. The phase thickness of the film for an angle of incidence ϕ_0 is given by

$$\delta_s = 2\pi\nu\, d(n_a^2 - n_0^2 \sin^2\phi_0)^{1/2} \quad \text{for the s-component}$$

and $$\delta_p = 2\pi\nu\, d n_a\left(1 - \frac{n_0^2}{n_b^2}\sin^2\phi_0\right)^{1/2} \quad \text{for the p-component.}$$

The case of the gyrotropic film, in which gyromagnetic or gyroelectric effects occur, is somewhat more involved. The basic observation in gyrotropic media (such as, e.g. magnetized ferromagnetic media), is that when plane-polarized light is incident on a slab of the material, the reflected or transmitted light is elliptically polarized, with the major axis of the ellipse inclined to the direction of incident polarization. Formally, this may be written into Maxwell's equation by substituting for the permeability or permittivity, which are scalars for isotropic media or symmetric tensors for non-gyrotropic anisotropic media, the skew-symmetric tensors:

$$[\mu] \equiv \begin{bmatrix} \mu & -ip\mu & 0 \\ ip\mu & \mu & 0 \\ 0 & 0 & \mu_0 \end{bmatrix} \qquad 6.25$$

or

$$[\varepsilon] \equiv \begin{bmatrix} \varepsilon & -iq\varepsilon & 0 \\ iq\varepsilon & \varepsilon & 0 \\ 0 & 0 & \varepsilon_0 \end{bmatrix} \qquad 6.26$$

The values of p and q, respectively the gyromagnetic and gyroelectric 'constants', are complex and frequency-dependent. Substitution of these expressions in Maxwell's equations show that such media are birefringent. This applies if *either* the permittivity is a tensor *or* the permeability. In fact if both are present and interaction terms were to be important, then there would be *three* indices corresponding to a given direction. In fact for known materials, the gyro-constants p and q are individually much less than unity, so interaction terms may safely be ignored.

The case of reflection at the boundary of a gyrotropic medium which is magnetized (or electrified) in any arbitrary direction is somewhat complicated and these problems are considered in terms of three special cases, viz: (fig. 6.17)

Polar case – magnetization $\perp$ surface of separation,

Longitudinal (or meridional) case – magnetization $\parallel$ surface and lying in the plane of incidence

Transverse (or equatorial) case – magnetization $\parallel$ surface and lying $\perp$ plane of incidence.

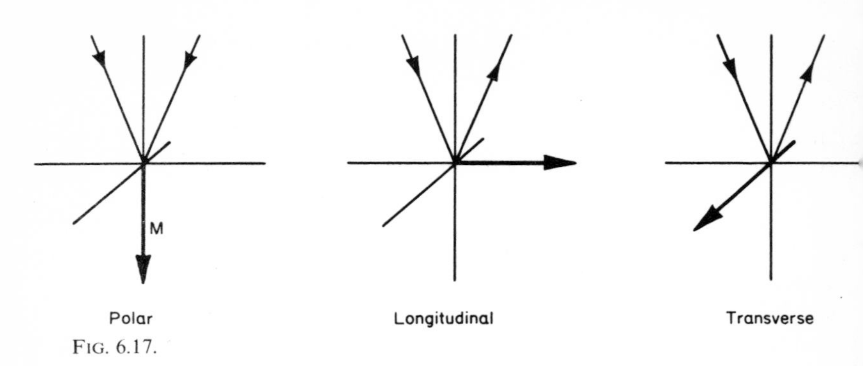

FIG. 6.17.

In dealing with gyrotropic media we lose the inherent simplicity of the isotropic case in which we may treat the p- and s-components independently. For an isotropic medium an incident p-component gives rise to refracted and transmitted rays in which the electric vector is confined to the plane of incidence. In the gyrotropic case, an s-component grows as the (initially) p-wave proceeds and the s- and p-components are no longer independent. The effect of this generalization is that the propagation matrix needed for a gyrotropic medium is (4×4) in place of the (2×2) which suffices for the independent s- and p-components in the isotropic case.

If we specify the total field components in the νth layer as E_x, H_y, E_y, H_x, then these may be written in terms of those in the next layer by the matrix relation

$$\begin{bmatrix} E_x \\ E_y \\ H_y \\ H_x \end{bmatrix} = (M) \begin{bmatrix} E_{\nu+1,x} \\ E_{\nu+1,y} \\ H_{\nu+1,y} \\ H_{\nu+1,x} \end{bmatrix} \qquad 6.27$$

where M is a 4×4 matrix involving the gyromagnetic constant, the phase thickness of the layer and the refractive indices.

The general case, without approximation is cumbersome but simplification results from noting that generally p and q are $\ll 1$ and that terms in p^x, q^x with $x > 2$ can be neglected. Thus for the polar case we obtain, for M:

$$(M) \equiv \begin{bmatrix} \cos\delta & \frac{i}{m}\sin\delta & \frac{1}{2}i\delta^2 p & 0 \\ im\sin\delta & \cos\delta & m\delta p & -\frac{1}{2}i\delta^2 p \\ -\frac{1}{2}i\delta^2 p & 0 & \cos\delta & -\frac{i}{m}\sin\delta \\ m\delta p & \frac{1}{2}i\delta^2 p & -im\sin\delta & \cos\delta \end{bmatrix} \qquad 6.28$$

where $\delta = 2\pi\nu nd\cos\phi_0$, $m = c/n$ and p is the gyromagnetic constant. Corresponding expressions may be derived for the longitudinal and transverse cases (See Smith, 1965).

One feature of interest in the behaviour of gyromagnetic media is that of the reflected beams in the polar case. It emerges that the refractive index of the medium for a wave travelling at an angle $+\theta$ with the polar axis is different from that for a ray at $-\theta$. Thus the law of reflection does not apply in this circumstance. In fact a single ray incident on a film *will* break into two on entering the film and the two waves couple in the manner illustrated by fig. 6.18.

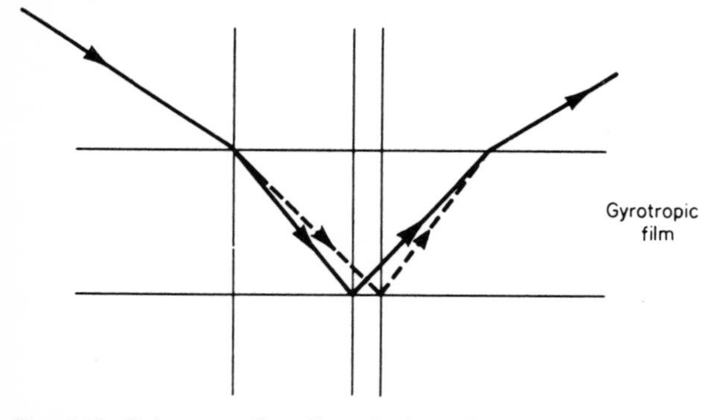

FIG. 6.18. Polar case. Coupling of refracted waves.

The detailed behaviour of a gyrotropic film may be calculated in terms of the film constants and gyrotropic parameters. Thus the ellipticity and rotation with respect to the incident beam may be determined. In principle, it is possible to decide whether a film's gyrotropy springs from gyromagnetic or gyrotropic effects or both, although, since most materials of interest are very heavily absorbing and the ellipticities produced are very small, the problem is generally somewhat difficult.

7

Magnetic Properties of Films

7.1 *General features of the magnetic behaviour of films*

In the preceding chapter, stress was laid on the dependence of the optical behaviour of thin films on their microstructure. Such comments apply perhaps with even greater force to the ferromagnetic behaviour of films. (When in this book we refer to 'magnetic' properties we imply 'ferromagnetic'. Para- and dia-magnetic effects in thin films are too small to be easily observable and are of limited interest anyway). As with bulk materials we ought to be able to distinguish between structure-dependent and structure-independent magnetic properties. Thus we could hope that quantities such as the saturation magnetization and Curie temperature would not seriously depend on structural features, while at the same time accepting that coercivity, remanence and micro-magnetic behaviour generally, will so depend. In fact we shall see that the situation is not so clear-cut as these remarks suggest.

We will start by considering, in the light of our knowledge of bulk ferromagnetism, just how we should expect the saturation magnetization of a thin film to vary with parameters such as thickness and temperature. One way of accounting for bulk ferromagnetism is in terms of a picture of interactions between spins on neighbouring atoms. We might feel intuitively that the extreme case of a film, namely a two-dimensional lattice, could well behave quite differently from its three-dimensional counterpart. Severe difficulties exist: on the experimental side, 'films' generally stubbornly refuse to organize themselves as perfect parallel-sided layers a few atoms thick. From the theoretical angle the many-body problem requires such drastic simplifications before it becomes tractable that a clear-cut answer on what we *should* expect is out of the question. The early

Heisenberg model of ferromagnetism assumed that magnetic effects arose entirely from interactions between spins of unpaired electrons in the d-shells of the ferromagnetic elements. It now appears that, in addition to this, exchange effects associated with the conduction electrons may also be involved. This complicates even further what is already a very difficult theoretical problem.

There have been two general approaches to the problem of calculating the magnetic properties of a film. One uses the Heisenberg model and the other the spin-wave method. These will now be discussed. It will be seen that (*i*) each is expected to give reliable results under certain conditions and (*ii*) judgement on their relative merits is hampered by uncertainties in the experimental results.

7.2 *Molecular field treatment*

The starting point of the Heisenberg treatment, developed by Valenta (1957) and by Corciovei (1961), is to consider the magnetization of an array of two-dimensional nets of spins. Magnetic effects are calculated on the basis of exchange interactions between nearest-neighbour atoms in a given plane together with those from nearest-neighbours in adjacent planes. The problem may be tackled at various levels of sophistication. In the simplest form, it is assumed that each atom carries the same spin energy and that only nearest-neighbour interactions need be taken into account. This leads to the result that, for an atom with z nearest-neighbours, spin alignment will disappear at a (Curie) temperature T_c given by

$$T_c = \frac{zJ}{2k} \tag{7.1}$$

where J is the exchange integral. On this basis any form of lattice would be ferromagnetic, with its Curie temperature increasing with an increase in the number of nearest-neighbours. The approximations are, however, severe and are relaxed somewhat in Heisenberg's second treatment in which it is arbitrarily assumed that, in place of spins of constant energy, the atoms carry spins with a Gaussian distribution giving the same mean energy as for constant spins. This leads to the result

$$T_c = 2J/k\left[1-\left(1-\frac{8}{z}\right)^{1/2}\right] \tag{7.2}$$

suggesting that a lattice would be ferromagnetic only if eight or more nearest-neighbours were present. Thus on this picture, the plane lattice would not be ferromagnetic. This is also the result of applying the Bethe-Peierls-Weiss method of studying order/disorder transformations, in which an exact solution is obtained for a group of near-neighbours and all the surrounding atoms are represented by an average field.

There is at present no means of knowing, from the experimental side, whether plane arrays of 'bulk ferromagnetic' atoms are ferromagnetic so there is no simple way of deciding which of the approximations discussed above is the more appropriate. It is, however, possible to calculate, using the above molecular field approach, the expected dependence on number of layers (and hence film thickness) of properties such as the Curie temperature and saturation magnetization. This has been done by Corciovei who obtains, for a pile (fig. 7.1) of n layers, each of N atoms, the following expression relating

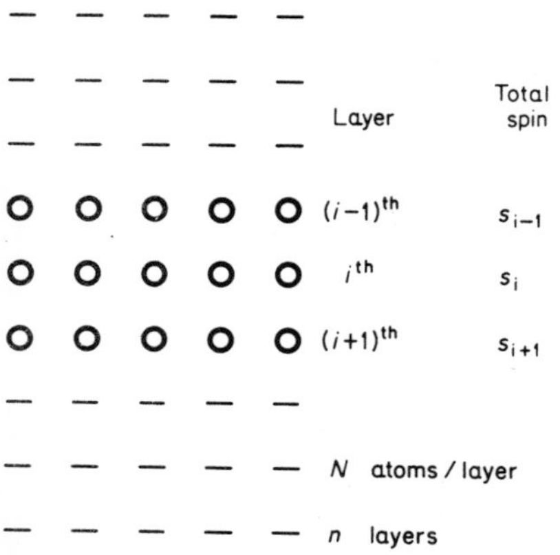

Layer | Total spin
$(i-1)^{th}$ | s_{i-1}
i^{th} | s_i
$(i+1)^{th}$ | s_{i+1}
N atoms / layer
n layers

FIG. 7.1.

the spin s_i of a given layer to the parameters of the system

$$s_i = \frac{N}{2}\tanh\left\{\frac{J}{NkT}\sum_{j=1}^{n} z_{ij}s_j + \frac{J^2}{2Nk^2T^2}\sum_{j=1}^{n}\left[-z_{ij}(s_i+s_j)+4z_{ij}\frac{s_i s_j^2}{N^2}\right]\right\}. \tag{7.3}$$

From this expression, the resultant magnetization of any number of layers n may be determined, as may also the temperature at which the magnetization vanishes. The expression for the Curie temperature reduces to that given in eqn 7.2. The dependence of Curie temperature and saturation magnetization of nickel on the number of layers is illustrated in fig. 7.2. These curves illustrate the experi-

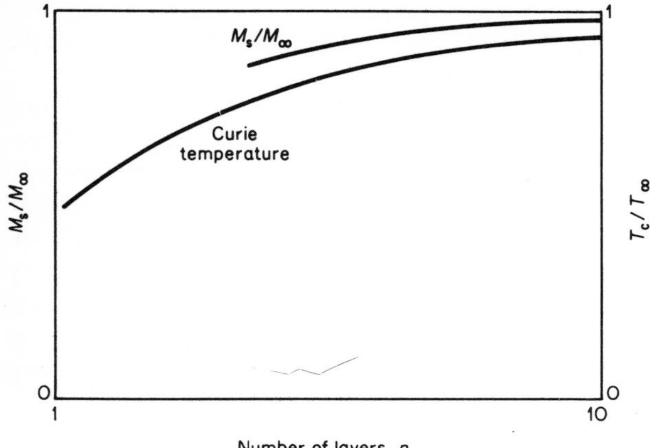

FIG. 7.2.

mental difficulty in verifying the predictions of the above theory. The saturation magnetization attains a value within a few per cent of the bulk figure for a film thickness of only ~1 nm. Since it is (*a*) difficult to make true films of this thickness and (*b*) difficult to measure the saturation magnetization of such thin specimens, then a close verification is not possible.

Experimental results will be discussed in Section 7.7 below. Before doing this, however, we shall examine a second theory of thin film magnetic behaviour, which approaches the problem in a quite different fashion.

7.3 *Spin-wave Theory*

If we imagine a ferromagnetic specimen to be placed in a uniform magnetic field at 0° K and an r.f. field to be applied, then we should expect all the individual spin-vectors on the atoms to precess uniformly about the field direction. (This will apply provided the specimen is of such size or shape that boundary effects are unimportant.) At temperatures above 0° K, or under conditions where boundary effects are important (as in a film) this will no longer apply. It proves possible to represent the magnetization in this case as a Fourier expansion of plane waves whereby each of the spatial harmonics is identified with a spin-wave. Although the reality of the situation is that of a set of discrete spins on a lattice, for large numbers of atoms (corresponding to a macroscopic specimen) the spin distributions can be described adequately in terms of (continuous) waves.

If the overall magnetization of a specimen with spins on a cubic array of spacing a is close to the saturation value, then the spin of a given atom may be written as

$$\mathbf{S} = \mathbf{S}_0 + \boldsymbol{\varepsilon} \tag{7.4}$$

where, for this case – which corresponds to temperatures well below the Curie point – $\boldsymbol{\varepsilon}$ is small compared with the unperturbed spin vector $\mathbf{S}_0$ By applying the usual cyclic boundary conditions appropriate to this type of problem, one finds that the spin-wave amplitude components $\boldsymbol{\varepsilon}_x$, $\boldsymbol{\varepsilon}_y$ take the form

$$\begin{aligned} \boldsymbol{\varepsilon}_x &= \varepsilon_0 \sin \omega t \sin \kappa_x x \sin \kappa_y y \sin \kappa_z z \\ \boldsymbol{\varepsilon}_y &= \varepsilon_0 \cos \omega t \sin \kappa_x x \sin \kappa_y y \sin \kappa_z z . \end{aligned} \tag{7.5}$$

The frequency and wave-vector of the spin waves are obtained from the equation of motion of the spin-vector and are related by:

$$\hbar\omega = 2\mathbf{S}_0 J a^2 \kappa^2 \tag{7.6}$$

where J is the usual exchange integral. For a rectangular box of volume V containing a cubic array of atoms the energy of a spin-wave of n quanta is given by

$$n\hbar\omega = \frac{J\kappa^2\varepsilon_0^2 V}{8a}. \tag{7.7}$$

If the notion of spin-waves is applied to the near-saturation case of a three-dimensional lattice, of spacing a the magnetization at temperature T, $M_s(T)$ is given, in terms of the 0° K saturation value, by

$$M_s(T) = M_s(0) - \frac{\mu_B}{\pi^2} \int_0^\infty \frac{\kappa^2 \mathrm{d}\kappa}{\mathrm{e}^{\quad /k} - 1} \tag{7.8}$$

where μ_B is the Bohr magneton. This leads to a $T^{3/2}$ law for the variation of $M_s(0) - M_s(T)$ with temperature (Bloch law).

In the application of spin-wave theory to films, Glass and Klein (1958) show that the saturation magnetization of a film with edges long compared with the atomic spacing is expected to depend on the number of atomic layers in the film. For a crystal of volume V containing N atoms in the form of G_z layers each consisting of square arrays containing $G \times G$ atoms ($G \gg 1$), the saturation magnetization at temperature T is given by

$$M_s(T) = \frac{N\mu_B}{V}\left\{1 - \frac{G^2}{N\pi}\sum_{n=0}^{G-1}\int_{2\pi/G}^{\infty}\frac{\kappa\,\mathrm{d}\kappa}{\exp\left[J/kT(\kappa^2 + 4\sin^2 \pi n_z/G_z)\right]}\right\}.$$

The form of the dependence of $M_s(T)/M_s(0)$ as a function of T/T_B, where T_B is the temperature at which the magnetization vanishes, is shown in fig. 7.3 for a face-centred-square film with 3×10^7 atoms

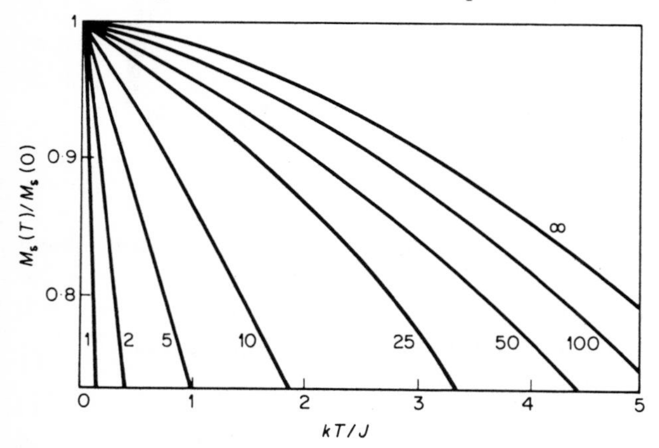

FIG. 7.3. Normalized saturation magnetization vs normalized temperature. Face-centred-cubic array, 3×10^7 units square. Figures on curves indicate thickness (as number of layers).

along the edge. The curves are given only for the range

$$0{\cdot}75 < M_s(T)/M_s(0) < 1$$

since the theory is not expected to apply to lower values of this ratio. It is seen that the saturation magnetization vs temperature curve for a film consisting of 25 layers (8·7 nm for nickel) differs significantly from that for an infinitely thick crystal, which may be taken to represent bulk material. The experimental results of

Neugebauer (1959) indicate no such marked dependence (fig. 7.4),

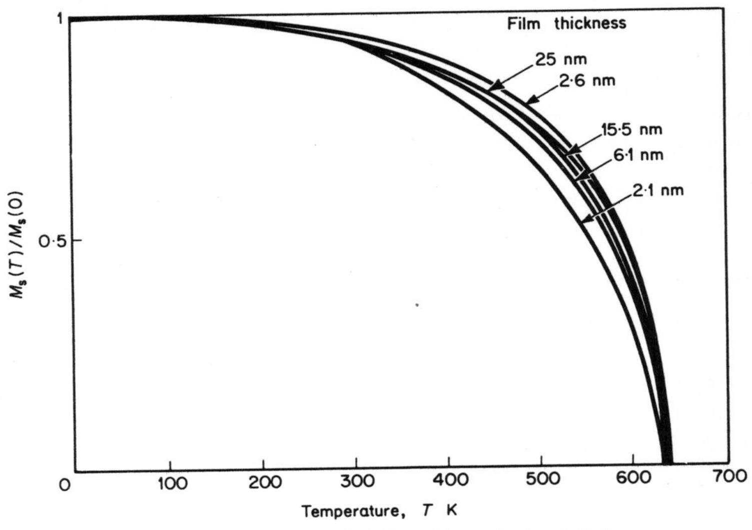

FIG. 7.4. Experimental results on nickel films. (Neugebauer, 1959).

certainly for films of thickness greater than about 3 nm. These measurements were made under vacuum at 10^{-9} torr; when the films were exposed to air, the saturation magnetization curve changed, especially for the thinnest films, indicating that the properties were being influenced by oxidation.

One of the questionable assumptions made in the derivation of the spin-wave theory discussed above is that involving the use of periodic boundary conditions. Although a reasonable assumption for a bulk lattice containing many atoms in all directions, the assumption is clearly suspect for a film of only a few layers thickness. We note that exchange interactions for spins on the surface occur only with spins below the surface so that the surface spins are subjected to a different spin environment from those in the interior. It seems reasonable to suppose that the movement of such spins will be more highly constrained than those inside, i.e. that a degree of 'pinning' of the spins could occur. In further developments of spin-wave theory, account is taken of the effect of such surface anisotropy. The results show that for a sufficiently strong surface anisotropy, a much smaller difference between bulk properties and film properties is expected.

The effects of surface anisotropy are strongly influenced by the presence of oxide on the surface of the film. This can arise from (*i*) the effects of strain, where the oxide forms epitaxially on the metal surface, (*ii*) the effects of surface roughness, which may be changed on oxidation and (*iii*) through exchange interactions with an anti-ferromagnetic oxide on the surface. Although there is a large amount of supporting evidence for the concept of spin-pinning, there is as yet insufficient data to enable quantitative agreement between theory and experiment to be obtained.

7.4 *Anisotropy in magnetic films*

In bulk single crystals of ferromagnetic materials, the ease with which the crystal may be magnetized depends on the crystallographic direction. Thus in the case of nickel crystals, the easiest direction is the (111) followed by the (110) and the (100). For iron the order is reversed and the (100) is the easiest direction. In the unmagnetized state, domains of the sample are magnetized along easy directions. On applying a field, the directions of individual domain magnetization change generally by domain wall displacement, and for moderate fields lie along the easy direction nearest to the field direction. On further increase in applied field, the magnetization vectors rotate away from the easy directions so as to lie closer to the field direction. The underlying cause of such magnetic anisotropy is to be found in the combination of the effects of the electrostatic field in the crystal, arising from the presence of the atoms on the lattice, and the coupling between the orbital motion of the electrons and their spins. The orbital motion of the electrons is influenced by the crystal field (depending on its symmetry and therefore on the crystal structure) and makes the spins aware of the crystal symmetry through the spin-orbit interaction.

From consideration of the range of the forces which give rise to anisotropy, there is no reason to believe that the magnetocrystalline anisotropy of a film of more than say a few tens of lattice spacings in thickness will differ significantly from the bulk *provided* the film has been prepared in zero magnetic field. When films are deposited in the presence of a field, they exhibit striking anisotropic properties.

Studies of this field-induced anisotropy have preoccupied workers for the last forty years or so. Much of the effort has been devoted to low-coercivity alloys of nickel and iron (Permalloy). In part, this is due to the fact that such alloys, the direction of whose magnetization can be switched by applying a very small field, are of great importance in information storage devices with rapid access.

Unravelling of the problem of film anisotropy ('the Permalloy problem') has proved extremely difficult. In part this is due to the bewildering array of factors which can influence anisotropy and in part because many of the early experiments were made on films produced in rather poor vacua, in the 10^{-5} torr region. Several hundred papers have been written on this subject over the years. For the most part they have led to a wider understanding of the complexity of the problem, rather than to any coherent solution.

7.5 *Theory of magnetic annealing*

If attention is fixed on the magnetic behaviour of a pure metal, then it is readily established that the effects of coupling between magnetic dipoles cannot, to a first order, account for the anisotropy observed. When, however, account is taken of second-order quantum mechanacal effects, then anisotropy is to be expected. If we consider a binary alloy, a different situation emerges, especially when we study the behaviour of such a system under conditions where diffusion can occur (e.g. at temperatures approaching the Curie point). Specifically, we examine the behaviour expected when a magnetic field is applied to a binary alloy held at a temperature such that diffusion can occur.

Experimentally such a process of magnetic annealing leads to the development of uniaxial magnetic anisotropy about the direction of the applied field. Suppose then that as a result of diffusion, two atoms of type A and B interchange positions. As seen in fig. 7.5, this results in an increase of one AA pair, one BB pair and a decrease of two AB pairs in the direction considered. Since the dipole-dipole energy will in general be different for the different pair combinations, the magnetic energy of the sample will be altered by this process. If the angle between the magnetization direction and that of the pairs is ϕ, then the change in energy associated with the diffusion process may be expressed in terms of ϕ and the coupling constants C_{AA}, C_{BB} and C_{AB}. This change of energy, for the case considered, is given by

$$\Delta W = (C_{AA}+C_{BB}-2C_{AB})B^2(1-\cos^2\phi) \qquad 7.10$$

where $$B = \frac{2S+1}{2}\coth\left(\frac{2S+1}{2}\cdot\frac{g\mu_B H}{kT}\right)-\frac{1}{2}\coth\frac{g\mu_B H}{kT} \qquad 7.11$$

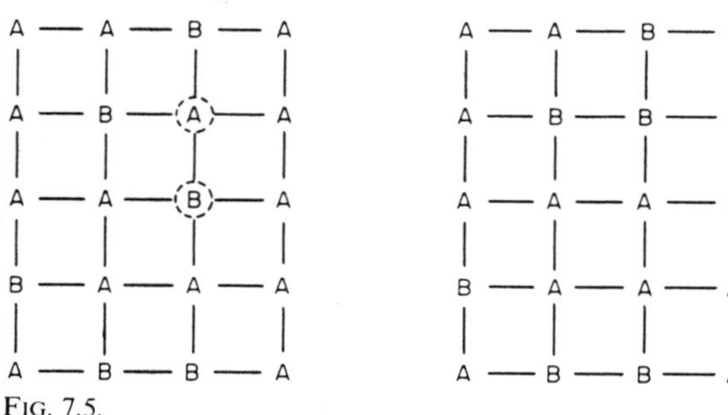

FIG. 7.5.

with the customary notation. Since ΔW is small compared with the energy of thermal agitation, the probability that a **BB** pair will form at an angle ϕ to the direction of the field is given by

$$p(\phi) = \frac{1}{n}\exp\{-(C_{AA}+C_{BB}-2C_{AB})B^2(1-\cos^2\phi)/kT\} \qquad 7.12$$

where n is the number of nearest-neighbours. Since, for different atomic constituents, C_{AA}, C_{BB} and C_{AB} will be different, it is clear that the influence of the applied magnetic field is to induce a structural, and therefore magnetic anisotropy in the crystal.

The above ideas are fully developed in theories by Néel (1953, 1954) and Taniguchi (1955) and the results compared with experimental findings on various nickel-iron alloys.

The general features of the theory are supported by experimental results inasmuch as the form of the angular dependence given by the theory is followed. More detailed predictions, such as the temperature-dependence of saturation magnetostriction constants on temperature and the concentration-dependence of the magnetic parameters of solid solutions, do not however yield detailed agreement with experiments on bulk materials.

Although it seems highly likely that the Néel-Taniguchi mechanism will operate in thin film specimens, as it appears to in bulk, it has become clear that this is only one of many factors which contribute to magnetic anisotropy of films. The existence of factors other than that due to pair-formation is shown in the experiments by Smith (1959) illustrated in fig. 7.6. The anisotropy of a range of

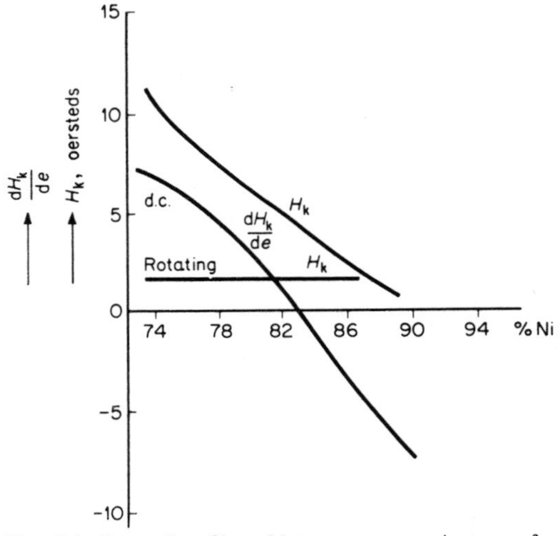

FIG. 7.6. Permalloy films. Note non-zero anisotropy for films formed in rotating field.

Permalloy films was studied, as a function of composition, in the neighbourhood of that for which the magnetostriction vanishes. Films were deposited (*a*) in a unidirectional steady field and (*b*) in a rotating field. In the latter case, no anisotropy would be expected if pair formation were the sole mechanism whereas, as seen in fig. 7.6, a constant anisotropy is observed. The decreasing anisotropy with iron content for the steady field case is consistent with the prediction of the pair-formation theory. However, when the observed dependence of anisotropy on temperature is compared with the theory, serious discrepancies are evident.

7.6 *Other sources of magnetic anisotropy*

From a survey of the structure and conditions of formation of films, there appear several possible mechanisms which could lead to anisotropy. The more obvious of these include:

(*i*) The influence of residual gas trapped in the film during deposition;
(*ii*) The effects of strain due to differing expansion coefficients of film and substrate;
(*iii*) The presence of structural anisotropies, due, e.g. to fibre orientation or to non-normal incidence of the condensing vapour stream;
(*iv*) The directional alignment of defects – vacancies, interstitials or dislocations.

In considering the influence of trapped residual gas, it is important to exclude the effects such gas may have on the overall magnetic properties of the film. These have been referred to earlier and are concerned with the effects on spin-pinning and surface anisotropy. Experiments carried out on nickel films deposited over a range of pressures from 10^{-10} to 10^{-5} torr reveal no significant variation of the film anisotropy. Similarly, no dependence is observed on the relative expansion coefficients of film and substrate (for films prepared under identical conditions) so that this too must be eliminated as a cause.

It is certainly true that the anisotropy observed in films deposited at non-normal incidence depends on the angle of incidence of the vapour beam. Electron microscopy of such films reveals a gross non-uniformity, as would be expected from the nature of the growth process. As individual crystallites in the film grow, they prevent the deposition of material in their shadow. This gross structural asymmetry leads to anisotropies in other film properties, e.g. optical and electrical behaviour. Thus although reasons can be advanced for the magnetic anisotropy of such structurally anisotropic films, this phenomenon throws no light on the anisotropy observed in uniform, normal-incidence films.

It appears most likely that the fourth mechanism suggested – involving the ordering of defects – will provide the most likely explanation of the phenomenon of magnetic anisotropy.

Direct experiments to test such a hypothesis are difficult to carry out – or even to imagine. However, considerable indirect evidence is accumulating from the study of the behaviour of films under annealing conditions. There remain, however, many unsolved problems.

7.7 *Domains in films*

The general features which give rise to domains in bulk ferromagnetic materials are well understood. In local crystalline regions, magnetization will lie along easy directions, in order to minimize the magnetocrystalline energy. In the neighbourhood of the boundary between regions magnetized in different directions, the spin directions do not change abruptly since this would involve a large spin-spin interaction energy. The directions of the spins change gradually on passing along the normal to the area separating the regions. If the resulting domain wall is thick, then spin-spin energy is low (since the angles between adjacent spins are small) but a substantial part of the material is magnetized along other than easy directions. The magnetocrystalline energy would then be high. The reverse applies to a very thin wall. The wall thickness is thus a compromise between (mainly) these two factors. There are many other factors – magnetoelastic effects, grain boundaries, shape effects and the like which complicate the detailed picture, but the general behaviour can be accounted for.

There are two main features through which films differ from the bulk in respect of domain behaviour. In the first place, the field-induced anisotropy, in films produced in a magnetic field, may well dominate so far as magnetocrystalline effects are concerned. Secondly, the limitation of specimen dimension in one direction means that, in the absence of an applied field, the magnetization will lie in the plane of the film on account of the large demagnetizing field which would result from any inclination to the film plane. Also whereas in bulk polycrystalline materials the crystal grain size is generally larger than the domain size, in thin films the reverse is usually the case. The shape inhibition of magnetization normal to the film plane also produces an alternative form of domain wall. Whereas in a bulk specimen, or a thick film, the spins rotate about the direction of the normal to the wall (Bloch wall) such a configuration in a very thin film would result in a large demagnetizing field at the wall centre. In this case the spin rotation will be about an axis perpendicular to the plane of the film (fig. 7.7) – a Néel wall. As would be expected,

Néel walls tend to be found in the thinnest films (e.g. up to about 300 Å for Permalloy) and Bloch walls for thicker films (say beyond 900 Å). In the intermediate thickness region, there is a tendency to

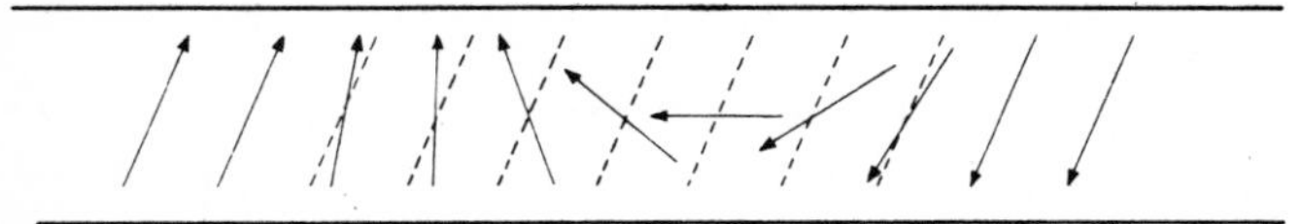

FIG. 7.7. Néel wall. The spins all lie in the same plane, parallel to the film plane.

form 'cross-tie' walls. The simple arrangement in fig. 7.8(a) will have a large stray field energy. This is reduced in the configuration of fig. 7.8(b) which, however, has a high spin-spin energy due to the adjacent reversed spins. The formation of a Bloch-type 'line' across

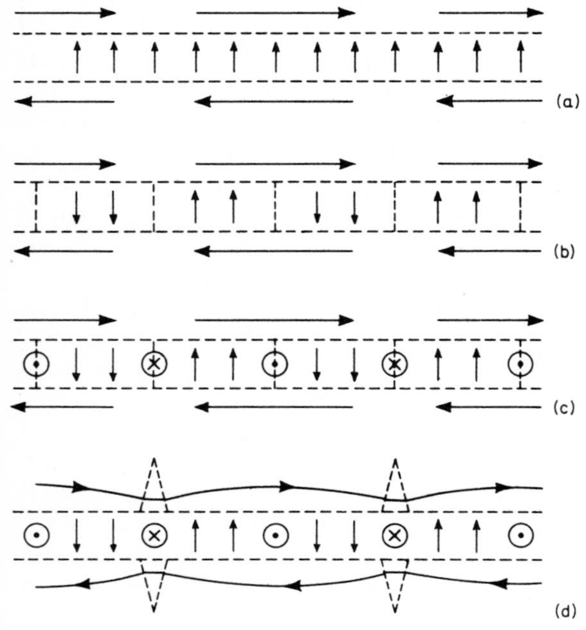

FIG. 7.8. Formation of cross-tie walls.

the wall (fig. 7.8(c)) reduces the spin-spin contribution but then leads to a spin distribution such that some flux closure is possible, with consequent lowering of stray field energy, by the development of transverse spikes as shown in fig. 7.8(d).

The general dependence of the energies of the different types of wall on film thickness is as shown in fig. 7.9 from a simplified calculation based on the assumption that the spin direction varies linearly

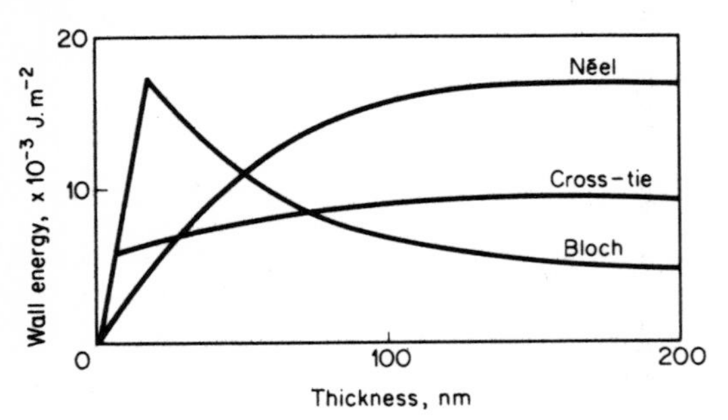

FIG. 7.9. Energies of different types of domain wall.

from one side of the wall to the other. (In fact, the calculation is not very sensitive to the precise form of spin variation.) The three types of domain wall referred to above can readily be distinguished by the usual Bitter pattern technique. The shape of the cross-tie walls is immediately obvious: Néel walls show more distinctly than Bloch walls and since, by the application of a magnetic field the type of wall may (except for the thinnest specimens) be changed, the wall type can easily be identified.

Although the above-mentioned types of wall are the most commonly observed, there is a variety of other kinds – spirals, concentric circles, ripples – which may occur under special conditions. These may be observed not only by the usual Bitter pattern technique, but also in the electron microscope, by the use of Lorentz microscopy. An electron passing through a specimen will, in the event of magnetization of the specimen, be subjected to the usual Lorentz force in regions where magnetic fields are present. Thus the residual fields in the neighbourhood of domain walls are clearly seen; electrons passing through these regions are lost from the main beam so that a weakened beam of electrons arrives at the image plane.

The use of Bitter patterns and Lorentz microscopy has led to a vast accumulation of information on the behaviour of domains and domain walls. As will be expected from the nature of the wall-forming process, the precise configuration expected will depend on specimen shape; on the mechanical strain in the sample, through the agency of magnetostriction; on the presence of imperfections or inclusions; and on the presence of surface anisotropy. Not unexpectedly, detailed analysis of the complex patterns frequently observed is often difficult. Although it overstates the case to suggest that detailed domain behaviour is fully understood, it is nevertheless true to say that even some of the most exotic domain patterns have been successfully accounted for on the basis of the Bloch and Néel concepts discussed above.

7.8 *Application of magnetic films*

Two properties of thin ferromagnetic films are such as to make them of especial interest. They are:

(*i*) The hysteresis loop may be almost perfectly rectangular and have a low value of coercivity, thus constituting a bi-stable device;
(*ii*) The magnetization may be switched from one direction to the opposite in an extremely short time interval. This is precisely the kind of device which is needed for the storage of information in binary code form.

The method by which such information storage has been effected in the early generations of electronic computer uses ferrite cores, which may be magnetized in one of two directions by a pulse of current passing through a loop surrounding the core. Such cores can be switched fairly rapidly (1 μsec) but the need for looping wires round them makes construction of vast arrays of them a cumbersome business. The potential offered by the thin magnetic film is illustrated in figs. 7.10 and 7.11. Figure 7.10(a) shows a typical rectangular hysteresis loop, such as is obtained for a Permalloy film deposited in the presence of a magnetic field. The loop of fig. 7.10(a) corresponds to switching in the direction in which the field was applied during deposition. In a perpendicular direction, the form of 'loop' is as shown in fig. 7.10(b). For the easy direction of magnetization of the film, we have the situation that the application of a field of less than $\pm H_c$ will produce no significant change in the magnetization of the film. For larger fields, the magnetization will be switched if the field is applied in the appropriate direction. If we imagine a set of films deposited on arrays of copper strips, as shown in fig. 7.11; and if the ferromagnetic films have been deposited in an applied field to yield an easy direction as shown, then binary information can be readily stored as right-to-left or left-to-right magnetization.

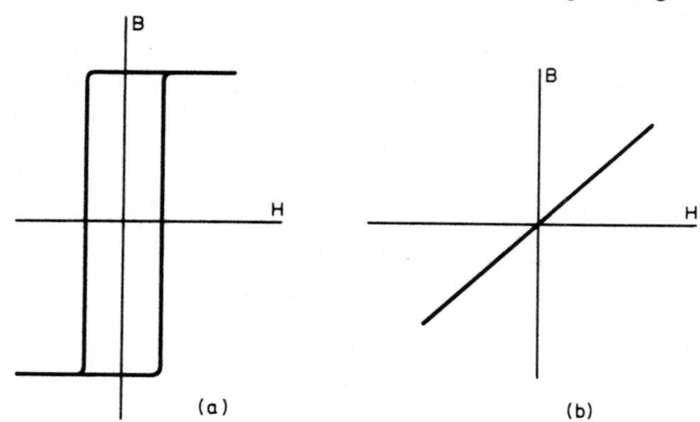

FIG. 7.10. (*a*) Magnetization curve of Permalloy – easy direction.
(*b*) Magnetization curve of Permalloy – hard direction.

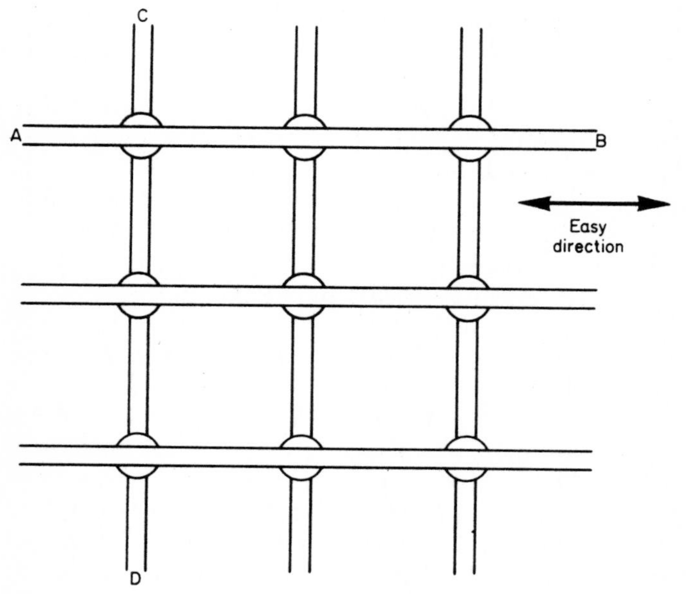

FIG. 7.11. Arrangement for a simple thin-film magnetic memory.

The method of writing in information is to apply a current pulse along AB which is large enough to hold the magnetization of the film in the direction perpendicular to the easy direction in the film. A pulse is applied along CD in such a way that the current in this pulse is still flowing when the current in AB falls to zero. When the AB pulse is removed, the magnetization of the film will relax to the easy direction. The direction in which it points will depend on the direction of the CD pulse. For the net shown in fig. 7.11, a current from C to D will produce a N-pole to the left. So long as the field produced by the currents in the CD strips is less than the coercive field, the directions of magnetization of all other films will not be changed by the AB/CD 'write' pulse combination.

The state of the magnetic stores on such an array can easily be determined by applying a pulse along one of the AB rows and by sensing the direction of the induced current pulses along the corresponding CD column. Thus if, as in fig. 7.12 a pulse is applied

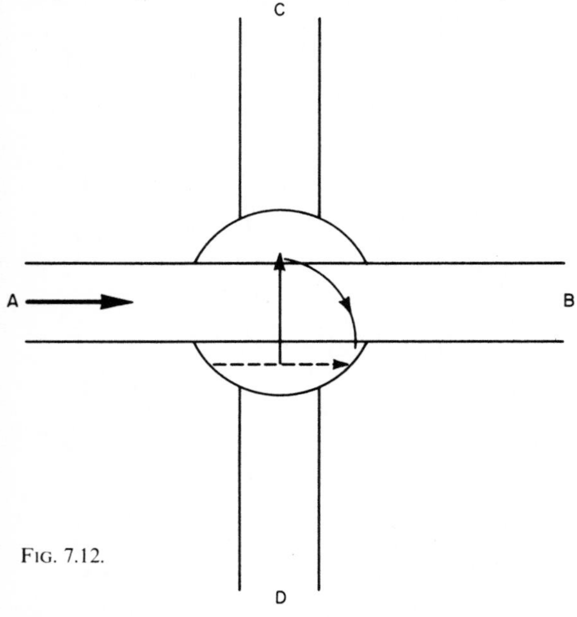

FIG. 7.12.

along AB, the magnetization – initially, say along DC – will swing as shown and will induce a transient current in the DC direction. For opposite magnetization, a reverse current would be recorded.

On removal of the AB 'read' pulse, the magnetization reverts to its initial direction. The read-out is non-destructive. Details of the many more elaborate arrangements in which thin film memory stores can be arranged need not concern us here. Suffice it to say that such stores offer many potential advantages over the older arrangements. A very large number of bits of information can be stored in a small space: the system of access strips and films can easily be made by conventional evaporation techniques: it is simple to arrange matters so that disturbance between different parts of the store is minimal: by the use of low coercive force materials, only low currents are required and the speed with which the magnetization can be switched is very high indeed. Thus whereas the switching time for the usual ferrite core is of the order of 1 microsecond (because the remagnetization involves a domain wall movement) that for a typical film system is of the order of a few nanoseconds.

8

Electrical Properties of Films

8.1 *General comments*

The electrical behaviour of films of all types of material – dielectric, semiconducting and metal – has become of particular interest in the period following the Second World War. Following the discovery of the transistor, it rapidly became clear that the conventional racks of valves, resistors and capacitors could in many cases be replaced by thin-film systems. The deposition of, e.g. thin metal films as resistive elements and of dielectric layers for capacitances could be easily and accurately controlled. Moreover, since such resistors are almost entirely surfaces, problems of heat-dissipation are relatively easily handled. Although there had for a long time been interest in the electrical behaviour of films arising from, e.g. the predominance of surface scattering processes over those in bulk materials, the technological interest produced a great increase in the effort devoted to this subject. At the same time the experimental techniques for film production and for study under rigorously controlled conditions underwent considerable improvements. Since electrical characteristics come under the heading of 'structure-sensitive' properties, this has been of the greatest importance. In fact, much of the very early work on the conductive properties of metal films is impossible to interpret owing to the absence of relevant information on conditions of formation, film structure, thermal history and other relevant factors.

A dramatic example of the sensitivity of conducting properties of metal films to environment is illustrated by the observations of Professor Mayer in his laboratories in Clausthal. Mayer was studying the resistance of films of nickel prepared and maintained in an ultra-high vacuum system. It was found that when a flask of liquid hydrogen was brought into the laboratory, the resistance of the nickel film *inside an ultra-high vacuum system at* 10^{-10} *torr* changed perceptibly. No change was observed on the pressure gauge connected to the system. Such an observation might well cast gloom and despondency on anyone struggling to establish sufficiently rigorously controlled conditions for reproducible measurements.

In this chapter we shall review the electrical properties of metallic, semiconducting and dielectric films and deal briefly with the superconducting properties of films. We shall examine the way in which simple limitation of one dimension is expected to modify film behaviour and also the fact that other mechanisms, e.g. trapping effects at film imperfections, may play a part in determining film properties. We shall conclude with a brief description of the way in which films systems may be built into microcircuits.

8.2 *Conductive properties of metal films*

The broad question which one can ask is 'can the observed electrical behaviour of films be satisfactorily accounted for simply in terms of the effects of limitation of one dimension'? The answer, for films which fulfil the requirement that only one dimension is small, i.e. for films which approximate geometrically to the ideal plane parallel-sided slab, is 'yes, to a reasonably high level of sphistication'. In other words, the understanding of the behaviour of films is generally as good as that of bulk metals. Even in the case of discontinuous films, such as are often found at small thickness values, the general features (arising from, e.g. three-dimensional limitation, intercrystallite tunnelling and even conduction via substrate) can be accounted for even though agreement on detail may be lacking. In some cases this arises from the need for an oversimplified model to represent the very complicated geometry of the discontinuous, polycrystalline film.

Let us first consider the way in which we should expect the conductive properties of a parallel-sided metal slab to depend on thickness. Taking the simplest classical view of a metal as an assembly of free electrons characterized by a phenomenological mean-free-path, it is clear that provided the film thickness is very large compared with the mean-free-path, the conductivity will be expected to be practically the same as that of a bulk sample of identical structure to that of the film. The effects of the surface will be to introduce additional scattering of electrons whose last collision occurred within one mean-free-path of the surface, measured along the direction of travel. For thick films, such additional scattering effects will constitute only a minute fraction of the total scattering which gives rise to resistance.

When the thickness (of a 'perfect' slab) is of the order of the electron mean-free-path, then clearly the role of surface scattering will become important, even to the point of dominating other such processes. Before, however, we can estimate the effect of surface scattering, we need to decide whether, to consider the extreme possibilities, the electron is (*a*) specularly reflected or (*b*) diffusely scattered at the surface. For a mathematically plane surface, we might expect intuitively that the former would apply; for a sufficiently rough surface, the latter would perhaps be more likely. It is certainly true that the simple theory of *either* specular *or* diffuse reflection is often not followed by real films and it has become customary to assume that a proportion p of the incident electrons is specularly reflected and the remainder diffusely reflected. The parameter p is then deduced from examination of the experimental results.

If we assume completely diffuse reflection, we may readily determine the effective mean-free-path of an electron in a film of thickness t where the bulk mean-free-path is λ_0. For the situation shown in fig. 8.1, the actual free path of an electron scattered at P will depend on the direction θ of the scattering relative to the axis and will have the values indicated. The average free path for all directions and all starting points z is given by

$$\bar{\lambda} = \frac{3t}{4} + \frac{t}{2}\ln\left(\frac{\lambda_0}{t}\right). \qquad 8.1$$

For a system of N free electrons per unit volume obeying Fermi-Dirac statistics and for an electron velocity v at the Fermi surface, the conductivity σ_0 is given by

$$\sigma_0 = Ne^2\lambda_0/mv \qquad 8.2$$

Hence the conductivity σ of a film of thickness t would, on this model, be given by

$$\frac{\sigma}{\sigma_0} = \frac{\bar{\lambda}}{\lambda_0} = \frac{3t}{4\lambda_0} + \frac{t}{2\lambda_0} \ln\left(\frac{\lambda_0}{t}\right) \tag{8.3}$$

and so

$$\sigma = \frac{Ne^2}{mv}\left[\frac{3t}{4\lambda_0} + \frac{t}{2\lambda_0} \ln \frac{\lambda_0}{t}\right]. \tag{8.4}$$

This simple approach cannot be expected to give more than a rough indication of the expected conductivity change since, among other things, it assumes that the electrons whose paths terminate on the

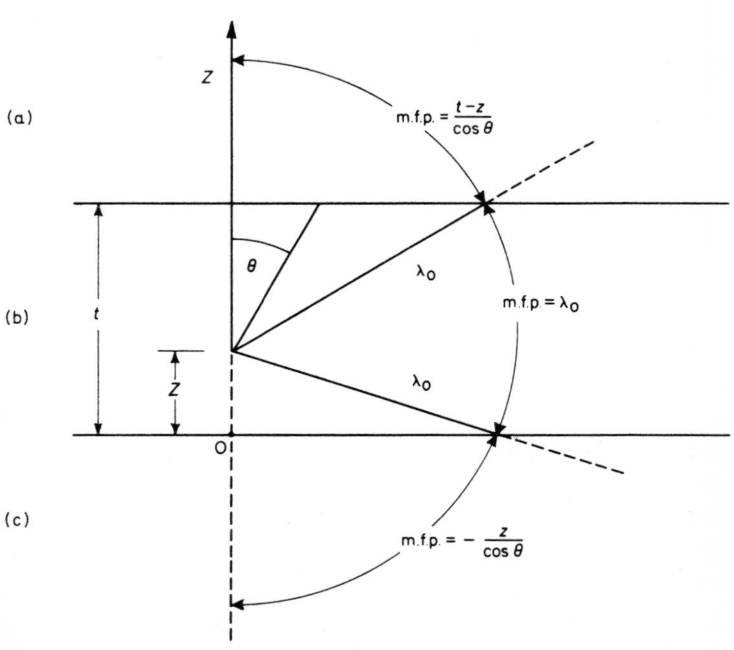

FIG. 8.1. Shows three regions. In region B, the electron travels a bulk m.f.p. In regions A and C, scattering at the surface occurs.

surface die on the spot and do not thereafter contribute. Moreover, it is clear that the right-hand side of eqn 8.4 does not approach unity as $t/\lambda_0 \to \infty$, as it should. Over the range $0 < t/\lambda_0 < 2$, the march of σ vs t/λ_0 is as shown in fig. 8.2, suggesting that the film conductivity might be ~75% of the bulk value when the film thickness is equal to the mean-free-path.

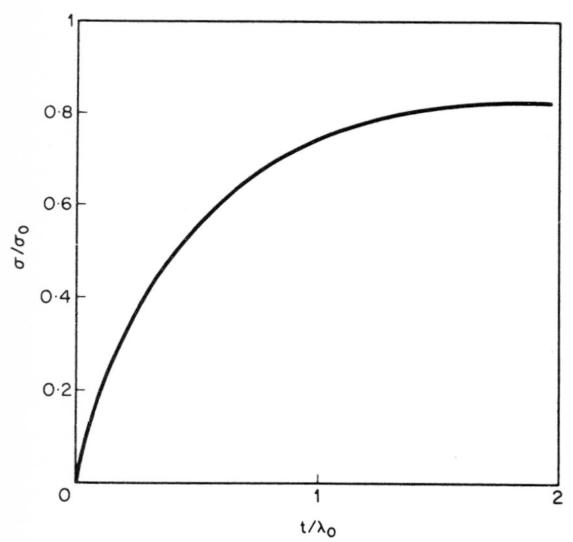

FIG. 8.2. Variation of σ with t/λ_0 for range $t \sim \lambda_0$.

The limitations of the above simple approach are removed in the treatment of Fuchs (1938) and Sondheimer (1952) who consider the general form of the solution of the Boltzmann equation for the case of a conducting film. In its general form, the Boltzmann equation gives the time- and space-dependence of the distribution function f under the combined effects of collisions and any external forces. Thus in equilibrium, we shall have

$$\left(\frac{\mathrm{d}f}{\mathrm{d}t}\right)_c = \left(\frac{\mathrm{d}f}{\mathrm{d}t}\right)_{\text{ext. forces}} = \frac{\partial f}{\partial t} + \vec{v}\,.\,\mathrm{grad}_{\bar{r}} f + \frac{\mathrm{d}v}{\mathrm{d}t}\,.\,\mathrm{grad}_{\bar{v}} f. \qquad 8.5$$

It is assumed that the distribution function will vary only in the direction normal to the plane of the film and that the collisional processes which give rise to electrical resistance effects can be characterized by a relaxation time. The source of the disturbance to the distribution is the applied electrical field E which, choosing for this the x-direction, produces an acceleration of $-eE/m$. If

$\left(\frac{\partial f}{\partial t}\right)_c$ is the rate of change of the distribution due to collision processes, then the Boltzmann equation for the thin film case, where the z-axis is normal to the film, is given by:

$$\frac{\mathrm{d}z}{\mathrm{d}t}\cdot\frac{\partial f}{\mathrm{d}z}-\frac{eE}{m}\cdot\frac{\partial f}{\partial \dot{x}}=\left(\frac{\partial f}{\partial t}\right)_c. \qquad 8.6$$

If collisional processes are represented by a relaxation time τ, then

$$\left(\frac{\partial f}{\partial t}\right)_c = -\frac{f(t)-f_0}{\tau} \qquad 8.7$$

where f_0 is the unperturbed distribution. The equation resulting from the combination of 8.6 and 8.7 may be solved by writing the distribution function in the form $f = f_0+f_i(v, z)$, where v is the velocity. The equation may be integrated by imposing the boundary conditions that, for the case of diffuse scattering, the distribution must be independent of the velocity with which the electrons leave the two film surfaces. This leads to a distribution of the form

$$f_i = \frac{eE\tau}{m}\cdot\frac{\partial f_0}{\partial \dot{x}}[1-a\exp(-z/\tau\dot{z})] \qquad 8.8$$

where $a = 1$ for electrons with velocity component in the positive z-direction and $a = \exp(d/\tau\dot{z})$ for negative-going electrons. With the distribution function known, the current density may be found and the conductivity of film and bulk compared. The result is

$$\sigma = \sigma_0\left\{1-\frac{3\lambda_0}{2t}\int_0^{\pi/2}[1-\exp(-t/\lambda_0\cos\theta)]\sin^3\theta\cos\theta\,\mathrm{d}\theta\right\}. \qquad 8.9$$

The integral cannot be evaluated explicitly but can be approximated for the cases $t \gg \lambda_0$. For thick films, we obtain

$$\sigma \fallingdotseq \sigma_0\ 1-\frac{3\lambda_0}{8t} \qquad 8.10$$

and for films of thickness small compared with the mean-free-path,

$$\sigma \fallingdotseq \sigma_0\cdot\frac{3t}{4\lambda_0}\left[\ln\frac{\lambda_0}{t}+0{\cdot}4228\right]. \qquad 8.11$$

The ranges of validity of the approximate formulae 8.10 and 8.11 compared with the computed (σ, t) dependence are shown in fig. 8.3.

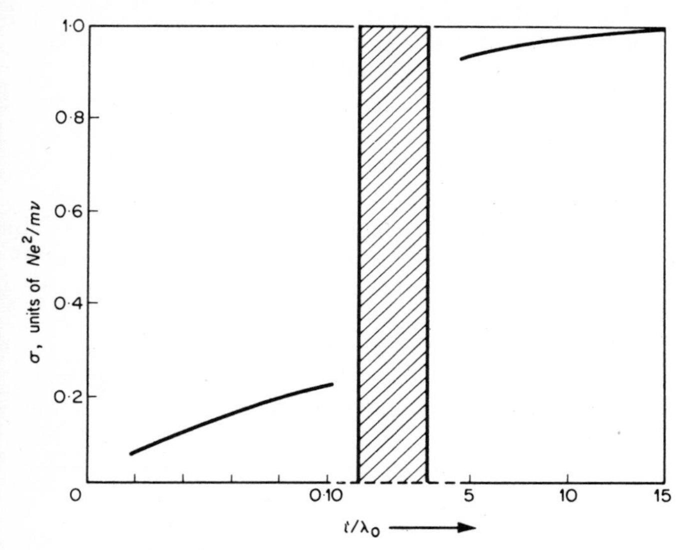

FIG. 8.3. Small- and large-thickness variation of σ with thickness. In the shaded region, neither approximation is valid.

In the above analysis, diffuse reflection of electrons at the film boundary has been assumed. When allowance is made for a fraction p of the incident electrons to be specularly reflected, the above expressions are modified. For thick films, equation 8.10 becomes

$$\sigma = \sigma_0 \left\{ 1 - \frac{3(1-p)\lambda_0}{8t} \right\} \qquad 8.12$$

whilst the expression for thin ($t \ll \lambda_0$) films takes the form

$$\sigma = \sigma_0 \cdot \frac{3t}{4\lambda_0}(1+2p)[\ln(\lambda_0/t) + 0{\cdot}4228] \qquad 8.13$$

for the case where $p^2 \ll 1$.

Clearly even this (somewhat empirical) elaboration may be insufficient to deal with the real case of a film on a substrate. In general, we should expect the specular/diffuse reflection ratio to be different for film/air and film/substrate interfaces. If it is assumed that the ratio is p at one surface and q at the other, then an integral corresponding to the development of that in equation 8.9 may be obtained. Numerical integration is required except for the limit of thick films, for which the value of p in 8.12 is replaced by $\frac{1}{2}(p+q)$. In fact, the availability of two disposable parameters has not led to any significant improvement in the agreement between experimental and theoretical results, in part because in most real films there are other factors which are not taken into account in the above treatment.

In addition to the thickness-dependence of film conductivity, the temperature coefficient of resistivity has been measured for many film materials. This may be related to the conductivity by a suitable extension of Matthiessen's rule for a bulk sample. This rule states that the resistivity of a metal may be written as the sum of a temperature-independent term accounting for impurity and defect scattering and a temperature-dependent term representing scattering by lattice phonons. The corresponding expression for a film would need to include a term representing scattering at the film boundaries. Will this be temperature-dependent? The answer depends on the value of film thickness in relation to mean-free-path. The latter decreases with rise in temperature so the fraction of the electrons from a given region of film which are scattered at the surface will change. However, if $t \gg \lambda_0$, surface scattering constitutes only a small fraction of the total so that the overall effect is small and the surface contribution can be regarded as temperature-independent. When t is of the order of, or less than, λ_0, the variation of surface scattering with temperature will become large.

For a film, Matthiessen's rule takes the form

$$\rho_F = \rho_L + \rho_D + \rho_S \qquad 8.14$$

where $\rho_L = \rho_{L(T)}$ represents lattice scattering, ρ_D covers defect and impurity scattering and ρ_S takes account of surface scattering. The corresponding expression for a bulk metal is

$$\rho_0 = \rho_L + \rho_D \qquad 8.15$$

if it is assumed that the defect structure of bulk and film are the same, which will not generally be true. If, however, the assumption that defect scattering is sensibly temperature-independent is true, then conclusions drawn about temperature coefficients of resistivity will be valid. The temperature coefficients of resistance for film and bulk are, respectively, $\frac{1}{\rho_0}\frac{d\rho_0}{dT} \equiv \alpha_0$ and $\frac{1}{\rho_F}\cdot\frac{d\rho_F}{dT} \equiv \alpha_F$. Since for a thick film, the surface scattering contribution is essentially independent of temperature, then $\frac{d\rho_F}{dT} = \frac{d\rho_L}{dT} = \frac{d\rho_0}{dT}$, so that (Holland and Siddall, 1953):

$$\alpha_F \rho_F = \alpha_0 \rho_0. \qquad 8.16$$

Thus from 8.12, we see that

$$\frac{\alpha_F}{\alpha_0} = \frac{\sigma_F}{\sigma_0} = 1 - \frac{3(1-p)\lambda_0}{8t}. \qquad 8.17$$

It must be emphasized that this result is expected to hold only for cases where t is appreciably larger than λ_0. Measurements on epitaxially grown single-crystal gold films suggest that t/λ_0 needs to be about 4–5. Above this value, a consistent fit is obtained between theory and experiment for a p-value of 0·8 (fig. 8.4). These results

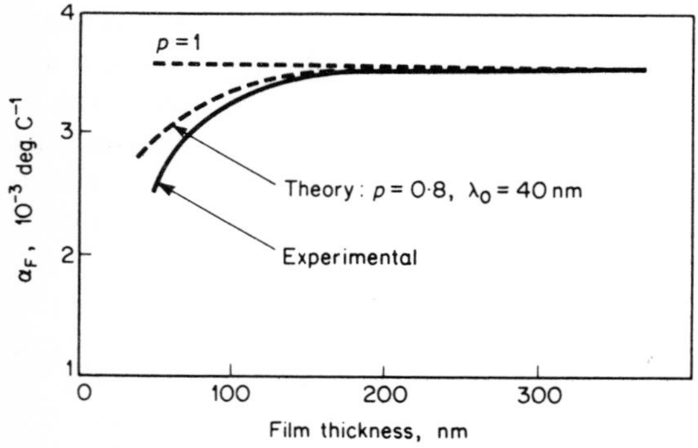

FIG. 8.4. Temperature coefficient of resistance vs thickness: epitaxial gold films.

are in contrast to those obtained on polycrystalline films, in which good agreement is obtained with the theory assuming diffuse scattering ($p = 0$). It seems certain that the difference in smoothness of the film boundaries is responsible for this effect. It is known that large areas of mica (on which the epitaxial films were grown) are atomically smooth and that under suitably controlled conditions films with atomically smooth surfaces may be grown. In fact it is possible under certain conditions to produce *polycrystalline* gold films with smooth surfaces, e.g. by sputtering on to bismuth or other metal oxides. In these cases, the electrical measurements agree with theory for large ($\sim 0{\cdot}9$–1) values of p. Similar behaviour is shown by smooth films of bismuth, lead, tin and silver.

As with the case of conductivity expressions, the equation for the value of α_F for any value of t/λ_0 is not explicitly soluble. For the case $t \ll \lambda_0$, the approximate solution

$$\alpha_F = \frac{\alpha_0}{\ln(\lambda_0/t) + 0{\cdot}4228} \qquad 8.18$$

applies. In practice, equation 8.18 gives a reasonably accurate result for $t \lesssim 0{\cdot}1\lambda_0$ and equation 8.17 for $t \gtrsim 2\lambda_0$. The fact that experimental results cannot be fitted, for thick films, for t-values below $4\lambda_0$ or so, suggests that the simple model, assuming identical scattering at the film/air and film/substrate surfaces, is inadequate.

It will be realized that the two extreme views – namely $p = 0$ and $p = 1$ – represent drastic simplifications and that the concept of a mixture of specular and diffuse reflection has more the flavour of a culinary recipe than a well-founded hypothesis. It *is* of course still a crude simplification, necessary in the face of the insoluble problem of exact calculation of the scattering behaviour of a real film surface.

It is by no means clear that a film with rough surfaces (where the rugosities may be on a scale comparable with the thickness) will be equivalent to a plane-sided film with an effective value of p between zero and unity.

Although this model is crude, it is in some respects remarkably successful. Consider, for example, the interpretation of the behaviour of bismuthfilms. By analogy with the behaviour of electromagnetic waves under total reflection conditions, the assumption has been made, for the electron-in-film case, that reflection is specular for angles of incidence above a critical angle and is diffuse for smaller angles. The agreement with the experimental results for bismuth is remarkably good.

In cases where experimentally produced films approximate reasonably to the parallel-sided isotropic slab, the results of electrical measurements are generally consistent with the mean-free-path theory discussed above. As mentioned in Chapter 4, the structures of very thin films of many materials bear not the slightest resemblance to the parallel-sided slab. It would not, therefore, be expected that the theory would apply to such discontinuous films. In general such films exhibit a negative temperature coefficient of resistivity and a complex dependence of electrical properties on external factors. Films of this type are discussed in the following section.

8.3 *The problem of discontinuous films* [Hill (1969)]

In the realm of ultra-thin 'films', the structure frequently takes the form of isolated crystallites on the substrate surface. How can such a film conduct? The mechanisms by which conduction might occur are (*a*) quantum mechanical tunnelling between individual particles, (*b*) thermionic emission of electrons into the conduction band of the substrate material, followed by capture by a neighbouring particle. If we consider particles of diameter p (e.g. hemispherical, although thus far, shape effects cannot be dealt with) separated by gaps of size g, then the relative values of p and g, together with their scale in relation to electron mean-free-paths in metal and substrate, will be expected to govern film behaviour. Clearly also, the type of substrate and its conduction mechanism will influence the overall behaviour of the system. In fact in cases where substrate behaviour plays a dominant role, we are studying the behaviour of the system film + substrate rather than simply that of the film itself. In these situations, variations of the resistivity of the film material (due to restricted size) may or may not be significant in determining the overall behaviour.

Suppose then that the particles are small and the gaps sufficiently small for quantum mechanical tunnelling to provide a dominant conduction mechanism. In such a case, the substrate plays only a supporting role and the field- and temperature-dependence of conductivity can be determined by the usual methods. In this situation, it is found that the conduction is field-dependent and leads to

$$\sigma \propto \exp(AV) \text{ for low fields}$$

and

$$\sigma \propto V^{1/2} \text{ for large fields.}$$

An activation energy will be associated with the conduction process. At low fields, this will arise from the existence of a barrier between adjacent charged particles. At high fields, the (normally Coulomb) potential distribution around the particles will be deformed by the field gradient, producing a local maximum, which determines the activation energy in this case.

If the separation between particles is increased, the probability of tunnelling will decrease rapidly until the contribution made by this mechanism is negligible. In this situation, conduction will occur through thermionic emission from the metal particles into the substrate. If the separation between particles is of the order of the mean-free-path of the electron in the substrate, then direct capture of the electron from one particle by the next, via the substrate, will occur. If, however, the separation is large compared with the free path, then the conduction will be determined mainly by the substrate conductivity.

For large particles separated by gaps small enough to give a high quantum mechanical tunnelling probability, the overall film resistance will be determined to some extent by the resistance of the individual particles. This will in turn be governed by the particle dimensions in relation to the electron mean-free-path and by the influence of defects, dislocations, grain boundaries and other imperfections in the crystallite. In contrast, in the case of large particles at large separations, the conductivity – via the substrate – will be so small that the resistive properties of the individual particles will be of no consequence.

It is seen, therefore, that an interpretation of the conducting properties of a discontinuous film even at a fixed temperature is a highly complicated problem, with many parameters which are often difficult to control or determine. An analysis of the temperature-dependence of conductivity brings in additional difficulties, e.g. from the variation in particle separation due to change of dimensions of the substrate with temperature. Thus the temperature-dependence will depend, in the worst case, on (*a*) the change in conductance of individual particles, (*b*) changes in tunnelling probability due to variation in crystallite separation, (*c*) variation in conducting properties of the substrate with temperature, (*d*) annealing effects in particle or substrate. Even if the effect of all these factors could be satisfactorily allowed for, one would in practice be beset by such factors as adsorption of residual gas in the 'vacuum' chamber, desorption of gases from the substrate or film or any of the multitude of uncertain and uncontrollable factors which influence the behaviour of films of this type.

One final point about the electrical properties of metallic films concerns the influence of a measuring current during deposition of the film. In general, the resistivity of a film increases above the bulk value as the thickness decreases and makes a large, abrupt increase at the stage where the film structure becomes discontinuous. The thickness at which this occurs is very much larger ($\sim$factor of two) for a film deposited with a measuring current flowing than for one without. This gross influence of measuring current on the film structure is amply confirmed by electron microscopy of films prepared in this way.

8.4 *Semiconductor films*

Early experiments on semiconductor films proved extremely difficult to interpret on account of the dominant influence of structural defects on the electrical properties. Thus mobilities were typically three orders of magnitude lower than those obtained on pure bulk single-crystals. In recent years, however, the greater control of deposition conditions and the use of epitaxy to produce monocrystalline films has led to a better understanding of the role of the factors influencing the film properties, although there remains much to be understood insofar as the structures of films depend on growth conditions which will, owing to the semiconductor's critical dependence on defects, affect the electrical behaviour. Thus it is known that, in the case of silicon films grown epitaxially on pure silicon, even the gross growth features depend on barely-observable amounts of impurity on the surface. Minute traces of carbon produce a nucleated, rather than a continuous, mode of growth.

Let us consider first the behaviour to be expected from a 'perfect' semiconductor film, regarded as a slab with bulk properties. We may perhaps ask whether the surface scattering treatment used for electrons in a metal film may be applied directly to the case of a semiconductor. Clearly in a semiconductor it is necessary to take account of space-charge effects at surfaces and to allow for the fact that semiconductors are characterized by long shielding length·. With these additional features included, the mean-free-path approach leads to predictions of the way in which film conductivity and mobility may be expected to depend on film thickness. (Zemel, 1966.)

Consider an intrinsic semiconductor for which the band structure at the surface is as shown in fig. 8.5. The potential at a distance z

from the film surface is represented by $\psi(z)$ and can be obtained in the usual way from Poisson's equation. For the potentials defined

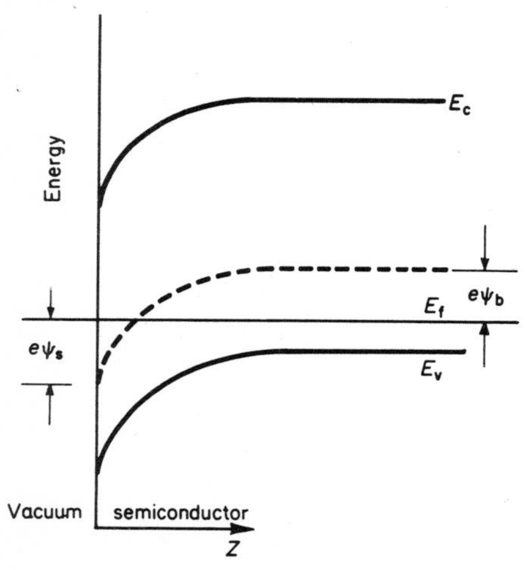

FIG. 8.5. Energy band diagram for semiconductor surface.

as in fig. 8.5, the charge density $\rho(z)$ is given, where $u \equiv [e\psi(z) - E_F]/kT$, by

$$\rho(z) = en_i\{\sinh u - \sinh u_b\} \qquad 8.19$$

which, subject to the boundary conditions $u \to u_b$ as $z \to \infty$ and $u = u_s$ at $z = 0$, leads to

$$E_z(z) = 2e\sqrt{\frac{n_i kT}{\varepsilon\varepsilon_0}}\{\cosh u - \cosh u_b + (u - u_b)\sinh u_b\}^{1/2} \qquad 8.20$$

where n_i is the carrier concentration and ε the permittivity. If we use the Boltzmann relation (equation 8.5), assume that collisional effects can be characterized by a relaxation time τ (equation 8.7) and apply boundary conditions on the assumption that the surface is the plane $z=0$, we can obtain computerized solutions for the distributions and hence obtain values for observable parameters such as the conductivity and mobility. The ratio of surface (u_s) to bulk (u_b) conductivity can be expressed as the integral

$$\frac{u_s}{u_b} = \frac{\displaystyle\int \frac{mv_x^2}{2kT} f_0(1-e^{-z/l})(e^{u_b-u}-1)\mathrm{d}v_x\mathrm{d}v_y\mathrm{d}v_z\mathrm{d}z}{\displaystyle\int f_0(e^{u_b-u}-1)\mathrm{d}v_x\mathrm{d}v_y\mathrm{d}v_z\mathrm{d}z}. \qquad 8.21$$

In the case of an intrinsic semiconductor, the integral may be obtained in a closed form if the ratio of the shielding length

$$L \equiv \left(\frac{\varepsilon\varepsilon_0 kT}{2e^2 n_i}\right)^{1/2}$$

to the mean-free-path is integral or half-integral. Although an artificial exercise, the solution in this form does enable the variation of surface mobility with the surface potential to be easily calculated.

For the extrinsic case the value of u_s/u_b can be obtained in the form

$$\frac{u_s}{u_b} = \frac{3L}{8l}\,\frac{\displaystyle\int_0^{u_s}\int_0^{\pi} e^{-z/l}\cos\theta\sin^3\theta\,\mathrm{d}\theta\,\mathrm{d}u}{(e^u - u_s - 1)}. \qquad 8.22$$

The uncertainties in the assumptions which are needed for evaluation of experimental parameters from the theory are such that detailed agreement with experiment is not obtained. However, the behaviour of the effects of surface scattering on mobility and of the dependence of mobility on temperature follow the same general pattern as that observed in the (few) experimental results obtained under controlled conditions. There are, however, many factors which cannot be as yet fully allowed for in the theoretical treatment of surface scattering. Among these are (*i*) the change in the phonon spectrum at a surface due to the existence of surface modes; (*ii*) the effects of strain due, in the case of films on surfaces, to the effects of differential thermal expansion; (*iii*) the influence of mosaic structure in films and (*iv*) the effect of more general imperfections such as dislocations, defects, non-stoichiometry and composition changes due to diffusion effects at the film-substrate boundary. Since in general the properties of crystalline semiconducting materials are very sensitive to impurities and imperfections, it is perhaps hardly surprising that the detailed understanding of their electrical properties is still a very long way off.

The problem of providing experimental results to illustrate the behaviour of semiconducting films is a severe one. Although, as indicated above, there are some features of the theoretical predictions which are generally followed, the variation between different experimenters' results – even on films of 'pure' material – is very considerable. The two semiconducting materials which are best understood in the bulk, silicon and germanium, are difficult to prepare in film form under completely controlled conditions. Their electrical behaviour is found to depend, in the case of thermally evaporated films, on the pressure, rate of deposition, nature of substrate, type of evaporation source and on the time after the completion of the deposition. (Yet strangely enough, in view of this apparent sensitivity to ambient factors, the properties of such films sometimes do not change when removed from the vacuum system and exposed to air.)

EPITAXIAL SILICON

We shall illustrate the behaviour of monocrystalline films of silicon by reference to the results of Itoh *et al.* (1969), who prepared (111) films epitaxially on crystals with a spinel structure. Simultaneously with the evaporation of the silicon, antimony was evaporated to produce doped, *n*-type films, with thicknesses of the order of 10 μm. At this order, one can expect essentially bulk behaviour inasmuch as surface scattering effects would not contribute greatly to the electrical properties. It was indeed found that at large ($\sim 10^{18}$ cm^{-3}) carrier concentrations, the Hall mobility of the films was almost as high as that of the corresponding bulk material. The situation was

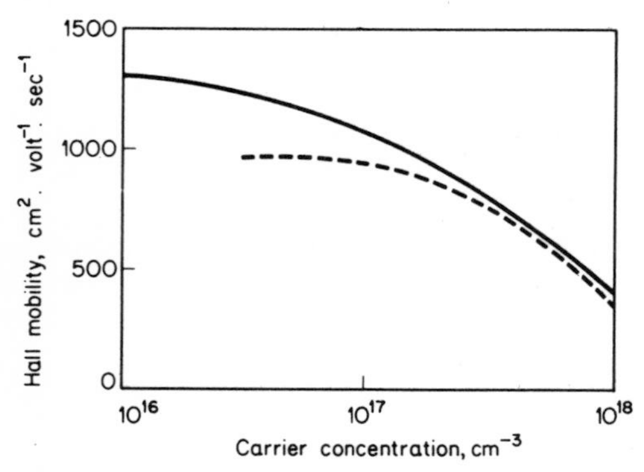

FIG. 8.6. Dependence of mobility on carrier concentration in epitaxial silicon films. (Full line: bulk Si; dotted line: Si film.)

less favourable at lower concentrations, as shown in fig. 8.6, in which the mobilities of film and bulk materials are compared.

The temperature-dependence of the Hall mobility depends markedly on the doping level, as shown in fig. 8.7. In the case of

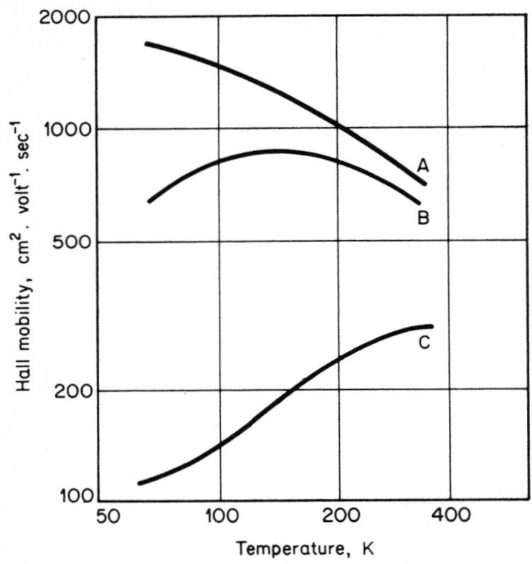

FIG. 8.7. Temperature-dependence of Hall mobility. Epitaxial silicon. A: light; B: intermediate; C: heavy doping.

heavily doped material, it is found that, over the temperature range indicated, the carrier concentration does not change significantly. This suggests that the impurity levels lie close to the bottom of the conduction band and are thus mostly ionized even at the lower end of the temperature range. It will be expected that scattering by impurities will be important at high doping levels. In lightly doped films, the carrier concentration is found to increase with rise in temperature over the range indicated, suggesting the presence of levels rather further below the bottom of the conduction band than in the former case.

AMORPHOUS SILICON AND GERMANIUM

In the case of amorphous films of these materials a somewhat different behaviour is observed. One immediate difficulty springs from the evolution of the conductivity of the films after deposition, presumably due to some form of structural annealing or to the diffusion of trapped gas. This effect can be large and is not understood. Thus an amorphous silicon film may have a resistivity of 0·3 Ωm on deposition, 0·5 Ωm an hour later and 1·5 Ωm after a few weeks. The dependence of electrical behaviour on the residual gas pressure in the deposition system appears to be confined to the region above $\sim 10^{-5}$ torr. Also, although low rates of deposition produce films of high resistivity, there is little variation (fig. 8.8) for deposition

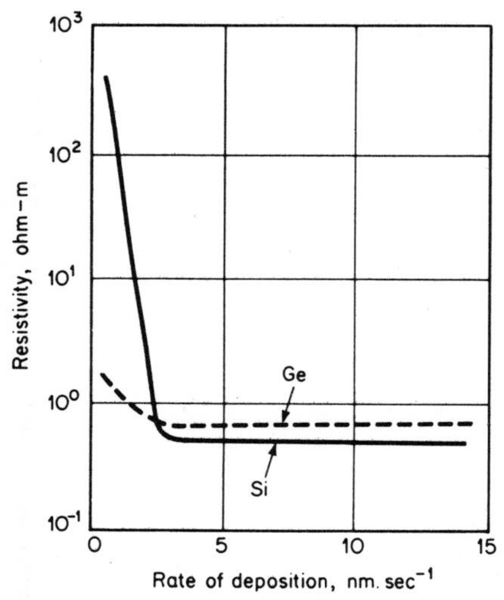

FIG. 8.8. Effect of rate of deposition: germanium and silicon films.

rates in excess of ~ 4 nm.sec^{-1}. One consoling feature of such amorphous layers is that the conducting properties tend to be dominated simply by the structural disorder so that the minute amounts of impurity which so dramatically change the behaviour of single-crystals have a rather smaller effect on amorphous films. Notwithstanding this favourable feature, large differences are found between different workers' results. Although some of these may be accounted for by differences in conditions of preparation (pressure, rate of deposition, etc.) differences by as much as a factor of 10^2–10^3 are sometimes found. One possible reason for discrepancies in the case of silicon concerns the possibility of the formation of an oxide during deposition. (Silicon is notoriously reactive.) Although small amounts of such impurity may not affect the material in the way that would hold for a single crystal, their presence may – by the evolution of crystalline and amorphous regions – affect the essential disorder on which the conducting properties depend. Thus the formation of minute crystalline regions in an otherwise amorphous matrix may provide trapping sites for charge carriers. If this were to be the case then at low applied fields, the conductivity would be essentially that due to the amorphous matrix. At high fields, trapped charges would be dislodged and jump from one trap site to another, thus giving a rapid increase of current with applied field. There is evidence, in the case of silicon and germanium, for a process of this general type since for fields above 10^6 volts.m^{-1} and at 77° K, the current/field dependence takes the form:

$$I \propto \exp(bE^{1/2}) \qquad 8.23$$

where b is a constant. Although the experimentally determined values of b differ (by a factor of 1·5–2) from those calculated on the basis of emission from trapped sites, the I vs E dependence has the expected form. The remarkable feature of the experimental results for many materials is the range over which plots of ln I vs $E^{1/2}$ are linear. A range of six orders of magnitude of current is not uncommon. Although close quantitative agreement with the predictions of trapping theories cannot be claimed, it does appear that this is the only plausible mechanism which could account for such results.

AMORPHOUS SILICON OXIDE

The electrical properties of this material are of considerable interest (and complexity!) in view of its use as a dielectric. The apparent vagueness in the description (silicon *oxide* – which oxide?) testifies to the fact that the composition of such films is somewhat uncertain. Thus depending on the conditions of preparation, this material can exhibit a wide range of refractive indices. Chemically, we can do little better than to call it Si_xO_y, knowing that when we evaporate the oxide in oxygen we obtain films with $x \simeq 1$ and $y \simeq 2$ whereas without the presence of oxygen, x and y are approximately unity. For this and other reasons, the conductive properties (for direct current) of silicon oxide films depend critically on all the usual deposition parameters, variation of which can produce conductivities ranging over factors of 10^8–10^9.

We must distinguish, as we did for amorphous Si and Ge, between low-field and high-field behaviour. It appears established that the low-field conductivity is ionic; conductivity due to sodium and hydrogen ions has been identified in this region. At high fields, electronic conduction occurs and it is possible to identify three regions of different behaviour. These are:

I: log(current) $\propto$ (field)$^{1/2}$; marked temperature dependence;
II: current $\propto$ (voltage)^n, with $n \simeq 1{\cdot}5$;
III: low-temperature, high-field region where, over range 4–50° K, electrical properties are independent of temperature.

The behaviour in region I is similar to that observed in amorphous Ge and Si at high fields and it appears that the mechanism of 'hopping' of electrons between traps, presumably from structural inhomogeneities, is occurring here. There is, however, evidence that under some conditions, quantum mechanical tunnelling of electrons between trapping centres may play a part. Also the injection of space-charge by the electrodes attached to the film may contribute. It seems likely, in fact, that the mechanism operating in region II is associated with space-charge effects of this kind. Region III is far less well understood, since the role of the electrodes is not clear. It appears possible that a combination of electrode effects and tunnelling phenomena is involved.

COMPOUND SEMICONDUCTORS

Compounds such as PbS, PbSe and PbTe are of considerable interest in view of their technological importance in devices such as infrared detectors, thermoelectric elements and microelectronic components. In many applications, these materials are conveniently used in thin film form and considerable effort has been expended in producing epitaxial layers. By way of example, the properties of epitaxial PbSe will be briefly discussed. Epitaxial films have been prepared using muscovite mica as a substrate. The mode of growth is by the formation of tetrahedral nuclei which, as the film thickness grows, coalesce to give a continuous layer, generally at thicknesses of the order of tenths of a micrometre. Replica micrographs of films of a thickness of $0{\cdot}5\ \mu$m reveal considerable surface texture.

The variation of Hall mobility with film thickness is of the form shown in fig. 8.9. The mobility for films of thickness 1 μm is found to be only $\sim 1/3$ of that for the bulk material. For this thickness, the observed low mobility cannot be accounted for by surface scat-

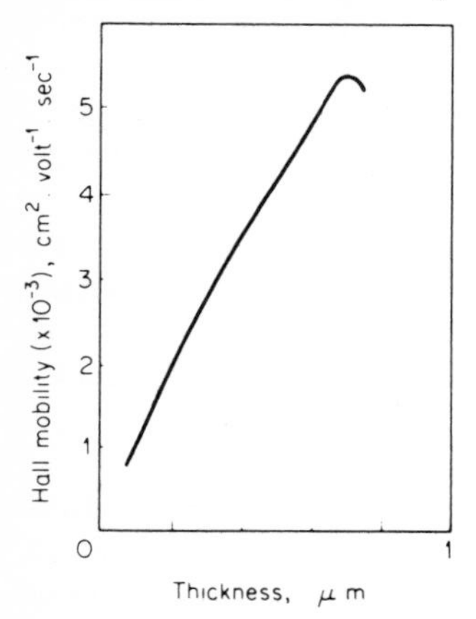

FIG. 8.9. Mobility/thickness relation for lead selenide films.

tering at the boundaries of a parallel-sided film. For such a film, the surface scattering would reduce the mobility by less than 5%. It is likely, in view of the mode of growth, that the films, although displaying electron diffraction spot patterns characteristic of monocrystalline layers, contain a high density of defects which act as scattering centres.

Although only a few examples of semiconducting films have been described, the general characteristics shown are fairly representative. On the whole, qualitative accounts can be given of the gross behaviours exhibited. In a few cases only is sufficient information available for a highly detailed description to be possible.

8.5 *Superconducting behaviour of thin films*

The importance of thin films in studies concerning superconductivity lies in the fact that when a superconducting specimen is placed in a magnetic field (of strength less than that which produces a transition to the normal state), the field strength within the specimen decays very rapidly with distance inward from the surface. Thus the behaviour of the uppermost layer (of thickness tens of nm) differs from that of the bulk. It would be expected, therefore, that measurements on thin films of superconductors could lead to information valuable for the development of theories of superconductivity and this is indeed the case.

A discussion in detail of the phenomena and theories of superconductivity would not be appropriate here and the reader is referred to 'Superconductivity' by A. E. Lynton. Some general comments relating to the development of this subject will serve to indicate the role and status of thin film studies in this field.

The onset of superconductivity occurs, for certain metals and alloys, when the temperature of the sample falls below a critical value. A current induced in such a specimen appears to persist indefinitely. From studies of the (lack of) decay of current in a superconducting circuit, it has been possible to set the upper limit of resistivity at around 10^{-25} ohm-m.

The phenomena exhibited by superconductors imposed interesting challenges on the existing theories of physics. At an early stage, it became clear that the observed behaviour could not be accounted for simply by inserting zero for the resistivity in the existing equations relating to the electrical behaviour of metals. For if this is done, the consequence of applying Maxwell's equations to the case of a conductor of zero resistivity is that the time rate of change of magnetic flux in such a perfect conductor is given by

$$\nabla^2 \dot{\mathbf{H}} = \frac{4\pi n e^2}{mc^2} \dot{\mathbf{H}} \qquad 8.24$$

If we consider a semi-infinite specimen, bounded by the plane $x = 0$, occupying the positive half-space, then equation 8.24 implies a variation with x of $\dot{\mathbf{H}}$ according to the relation

$$\dot{\mathbf{H}}(x) = \dot{\mathbf{H}}(0) \exp(-x/\lambda) \qquad 8.25$$

where $\lambda = (mc^2/4\pi n e^2)^{1/2}$. n is the number of conducting electrons per unit volume. For a typical metal, n will be of the order of the number of atoms per unit volume and, from the above, λ is seen to be of the order of 10 nm.

The implication of equation 8.25 is clear. At depths large compared with λ, $\dot{\mathbf{H}}(x) \to 0$, If, then, a normal specimen in a magnetic field is somehow turned into a 'perfect' conductor, any magnetic flux in the specimen at the instant of the transition will remain locked in the specimen after the transition has occurred. In a superconductor, however, the magnetic flux inside the specimen in the superconducting state is found to be zero (at depths larger than $x \sim 10$ nm). If magnetic flux is present when the transition normal → superconducting takes place, the flux is expelled to the exterior of the specimen. Thus the explanation of superconductivity requires something more sophisticated than the simple vanishing of electrical resistivity.

The fact that the presence of field variations (due simply to the presence of atoms) in a solid has no influence whatever on the behaviour of conduction electrons in the superconductor suggests immediately that the wave-functions describing the conduction electrons must be very different from those of the electrons in the normal metal. There clearly cannot exist any significant localization since this must result in scattering at the fields of the lattice atoms. If, for an infinite specimen, the electronic wavefunctions were of infinite spatial extent, then scattering at lattice fields would be vanishingly improbable. What is the implication of a wavefunction of infinite extent? From the uncertainty principle, this implies a precisely defined momentum. Now if an electron moves in a field **H** with velocity **v**, the momentum **p** is given by

$$\mathbf{p} = m\mathbf{v} + \frac{e\mathbf{A}}{c} \qquad 8.26$$

where **A** is the vector potential of the magnetic field. If the current flowing is **J**, then

$$\mathbf{p} = \frac{m}{ne}\cdot\mathbf{J} + \frac{e}{c}\cdot\mathbf{A} \qquad 8.27$$

$$= \frac{e}{c}\left\{\frac{4\pi\lambda^2}{c}\mathbf{J} + \mathbf{A}\right\} \qquad 8.28$$

where $\lambda^2 = mc^2/4\pi ne^2$ as before. Thus if any electrons (or current carriers) are in the state where **p** is *exactly* zero, the corresponding wavefunction will be spatially infinite. Equation 8.28 implies that, for this case,

$$\frac{4\pi\lambda^2}{c}\cdot\mathbf{J} + \mathbf{A} = 0 \qquad 8.29$$

the consequence of which, from Maxwell's equations, is that

$$\nabla^2\mathbf{H} = \frac{4\pi ne^2}{mc^2}\mathbf{H} = \frac{1}{\lambda^2}\cdot\mathbf{H} \qquad 8.30$$

instead of equation 8.24 for the time-derivative of **H**. Thus two results emerge, namely: (*i*) zero magnetic field is predicted inside the superconductor and (*ii*) the result 8.25 does not apply so that the field initially present at the normal → superconducting transition can be expelled.

A consequence of the above ideas, developed by London (1935) is that for the case of a superconducting ring (or for any multiply-connected superconductor) the flux through the hole due to the presence of superconducting carriers is quantized, in units of hc/q, where q is the charge on the carriers. Such flux quantization has been observed, the value of the flux quantum being such as to show that the carrier has a charge $2 \times$ (electronic charge) and not simply e, suggesting that *pairs* of electrons are involved.

Although a direct measurement of λ would serve to provide verification (or otherwise) of theories of superconductivity, it is difficult to interpret the results of experiments without knowing how the field decays within the specimen. It is, however, possible to study the variation of λ with parameters such as temperature, impurity content or applied field, without knowing the precise form of the penetration law. By the use of thin films of tin, lead and indium, studies of the variation of λ with temperature have been made (Lock, 1951). On the basis of the London equations, the expected temperature-dependence of λ should follow the relation

$$\lambda(T) = \lambda(0)\left[1-\left(\frac{T}{T_c}\right)^4\right]^{1/2} \qquad 8.31$$

where T_c is the superconducting transition temperature. Although this relation is followed over a reasonable range of temperature, there is an unmistakable departure at temperatures below about $0{\cdot}8\,T_c$. Moreover, other experiments reveal that the penetration depth of certain dilute alloys varies with the electron mean-free-path, although there is no reason why this should be so, from the derivation of the London equations.

These and other difficulties led to the emergence of modified treatments by Pippard (1950) and Ginzburg and Landau (1950) in which it was assumed that the wave-functions of the superconducting electrons are coherent only over a finite range. This relaxation of the rigid picture on which the London model rests leads to qualitatively satisfactory dependence of the penetration depth on the experimental parameters mentioned above. Thus it is observed that the critical magnetic field required to quench superconductivity increases as the thickness of the specimen decreases. For a film of thickness t, the ratio of the quenching field H_s for the film to the value H_c for a bulk specimen is given, by the Ginzburg-Landau theory, as

$$\frac{H_s}{H_c} = 2\sqrt{6}\lambda/t \qquad 8.32$$

where λ is the London penetration depth and it is assumed that $t \ll \lambda$. In fact in this and many other results of the phenomenological variations of the London theory, a closer fit is obtained if the observed penetration depth is used in place of λ.

The next stage in the development of a theory of superconductivity arose from, among other things, the observation that although the d.c. and radio-frequency resistivity of superconductors become zero below a critical temperature, the optical reflectivity showed no change. Vanishing of resistance was observed at frequencies of up to ≈ 10 Gc.sec^{-1} whereas for wavelengths below ≈ 30 μm (frequency 10^{13} c.sec^{-1}) normal metallic reflection was observed. This is suggestive of the existence of an energy gap in the electron spectrum. Quanta with energy less than that of the gap will be unable to excite electrons to the higher level and so no absorption will occur at these frequencies. The size of the gap can clearly be determined simply by measuring the variation with wavelength of the absorption of the superconducting film. In fact the absorption edge for all superconductors lies in the most difficult spectral region – the no-man's-land between the region of high-resolution optical spectroscopy (say out to wavelengths of the order of tens of micrometres) and the region of the klystron (millimetre wavelengths). Richards and Tinkham (1960) made such measurement by reflecting an infrared beam at several surfaces of a superconductor after which the energy of the final beam was measured by a bolometer. A typical result for a film of niobium is shown in fig. 8.10, where the fractional difference in power $\frac{P_s - P}{P}$, for the superconducting and normal states is plotted against wavenumber. It is clear that for wavenumbers above 30 cm^{-1}, there is no change on transition to the superconducting state. In other experiments, the transmittance of thin films, both in the normal and superconducting states have been measured over a wavelength range in the neighbourhood of the gap edge. The results may be fitted to an electromagnetic theory dispersion curve, characterized by a complex conductance $\sigma_3 = \sigma_1 - i\sigma_2$, where both σ_1 and

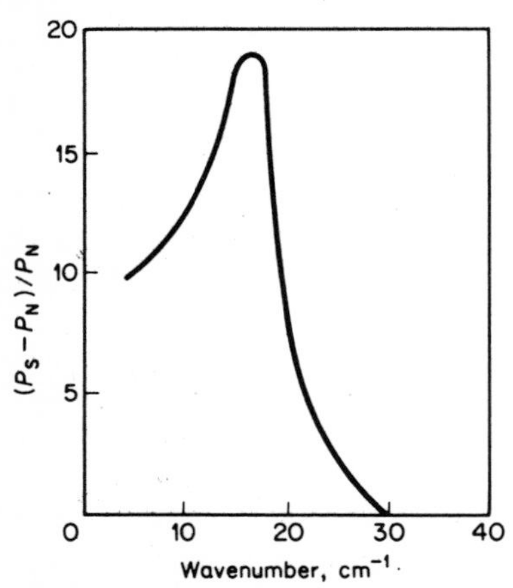

FIG. 8.10. Reflectance vs frequency for normal and superconducting niobium films.

σ_2 are frequency-dependent. The vanishing of resistive losses at low frequencies indicates that $\sigma_1 = 0$ up to a critical frequency, beyond which it is found experimentally – for all superconductors – that

$$\sigma_2 = \frac{\sigma_n}{a} \frac{kT_c}{\hbar\omega} \qquad 8.33$$

with $a = 0{\cdot}27$. This is precisely the result emerging from Pippard's theory. Moreover, the phenomenological penetration depth which, through Pippard's theory, can be evaluated from the experimental thin film results, is found to be several times the London value (equation 8.25) – a result which agrees with the various experimental determinations of this quantity.

BARDEEN, COOPER, SCHRIEFFER THEORY (1957)

Among the various results discussed above, there are two in particular which have led to the development of a satisfactory microscopic theory of superconductivity. One is the experimental demonstration of an energy gap, as revealed by the infrared transmission measurements on superconducting films and the other is the observation of *quantized* magnetic flux in which the energy of the quantum corresponds to the involvement of *two*, rather than of a single electron. Coupled with these results is the general observation that the phenomenon of superconductivity is patently *not* associated with one particular type of crystal structure, showing that the effect does not depend on one particular crystal field configuration. Furthermore, the superconducting transition is found to be extremely sharp, suggesting that co-operative effects involving very large numbers of current carriers are involved. Finally it should be noted that the energy difference between the normal and superconducting states is found to be very small indeed. One is thus searching for an electron-electron interaction with which a small energy is associated. Such a coupling is to be found in which lattice phonons are involved, in which a phonon produced by the scattering of one electron interacts with a second electron. It had been shown, in 1956 by Cooper, that an energy state lower than that for a normal metal could exist if electron states in the neighbourhood of the Fermi surface were filled with pairs of electrons with opposite spin and momentum. The idea of 'Cooper pairs' then led to the emergence of the BCS theory of superconductivity.

The general predictions of the BCS theory so far as thin films are concerned are essentially similar to those of the Ginzberg-Landau treatment. It is possible to relate the energy gap of the BCS theory to the order parameter of the Ginzburg-Landau theory and to deduce, *inter alia*, the dependence of energy gap on film thickness, field and temperature. It is expected, on the basis of the theory, that different behaviours would obtain depending on whether the film thickness is very small compared with the penetration depth or not. Thus for very thin films, the superconducting transition predicted is a second-order one, with no latent heat and without any discontinuity in the entropy. For thick films, a first order transition is predicted. The critical thickness above which the first-order transition is expected is given by $t_c = \sqrt{5}\lambda$ and this prediction is confirmed experimentally. Concerning the magnetic field-dependence of the energy gap, it is expected that this will be finite only for first-order transitions. Although explicit solutions of the Ginzburg-Landau equations for the thickness-dependence of the order parameter ($\propto$energy gap) cannot be obtained, numerical solutions show reasonable agreement with experimental results, as shown in fig. 8.11.

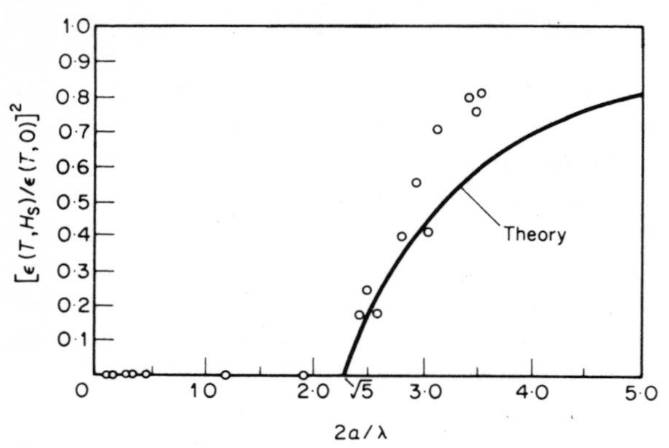

FIG. 8.11. Experimental results and the Ginzburg-Landau theory.

The critical magnetic field H_c needed to suppress superconductivity is generally found to be higher for thin films than for the corresponding bulk material. Thus in the case of films of tin, the dependence of H_c on temperature is as shown in fig. 8.12 (Zavaritskii, 1951). These films were deposited by evaporation on to glass and

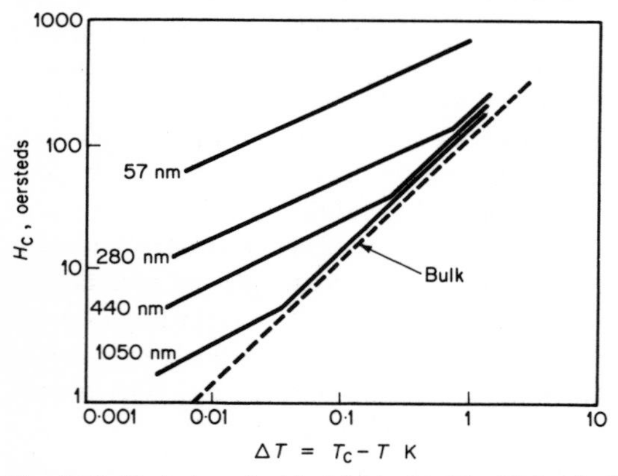

FIG. 8.12. Variation of critical field with $(T_c - T)$ for tin films.

were annealed at room temperature. Essentially similar behaviour is observed with films of thallium and mercury. The results for thicker films lie clearly in two parts, for one of which the slope of

$\ln H_c$ vs $\ln \Delta T$ is less than that of the bulk and for the other, the slope is the same as for the massive metal. The prediction of the Ginzburg-Landau theory (and also on the BCS theory) is that the slope of $\ln H_c$ vs $\ln \Delta T$ depends on the relative magnitude of the film thickness t and penetration depth λ. For $t \ll \lambda$, the form of the H_c vs ΔT relation is

$$H_c = A(\Delta T)^{1/2} \tag{8.34}$$

where A is a constant involving the energy difference between normal and superconducting states, the London penetration depth at absolute zero and the film thickness. The part of the line (fig. 8.12) to the left of the break has the expected slope of $\frac{1}{2}$. For the case $t \gg \lambda$, the variation of H_c with ΔT takes the form

$$H_c = B\Delta T + C(\Delta T)^{1/2} \tag{8.35}$$

where B and C are constants. The right-hand part of the line in fig. 8.12 can be fitted to an equation of this form. Somewhat curiously, the break is abrupt, although the theoretical curves, in the light of the approximations $t \gg \lambda$ or $\ll \lambda$, would be expected to apply accurately only *away* from the transition point.

There remain many features of thin film superconducting behaviour for which no satisfactory description is to hand. The indications are, however, that the essential ideas underlying the application of existing theories for bulk superconductors are capable of predicting the general features of thin film behaviour.

8.6 *Dielectric properties of films*

Interest in the dielectric properties of films stems in part from the fact that thin film technology offers the feasibility of producing very large capacitances in a small volume. Capacitance varies inversely as dielectric thickness and mechanical problems prevent the use of 'bulk' layers of dielectric with thickness much below the order of tens of microns. By many of the deposition techniques described in Chapter 2, controllable thicknesses down to tens of nm can readily be produced. If such films possess the same properties as those of bulk dielectrics, very large capacitances are attainable.

Somewhat frustratingly, the materials whose dielectric properties are best understood (although this should not be taken to imply a great depth of understanding) are those which are of little use as practical capacitors. The alkali halides have been studied extensively in many ways and much is known about, e.g. their structure, the nature of defects which form, the variation of their properties with temperature, etc. However, the fact that these materials are all soluble in water makes them of limited usefulness in electronic components.

In this section we shall discuss briefly the dielectric properties of alkali halide films, together with the current interpretation of their behaviour. We shall also indicate the general features of some of the materials which are proving of importance as practical capacitor materials.

ALKALI HALIDE FILMS

The real and imaginary parts of the dielectric constant of bulk alkali halides are found to vary with frequency in ways which can generally be understood. At low frequencies of applied field, such that the motion of the ions in the crystal can follow the field direction with negligible phase lag, there is a large ionic contribution to the polarizability. At the other extreme of high frequencies, the inertia of the ions prevents them from following the field reversals so that an electronic contribution to the polarizability is observed. In this region, the permittivity of the crystal should be (and generally is) equal to the square of the optical refractive index. Characteristic absorption bands occur in the infrared region (restrahlen bands) due to the energy dissipated when, under the influence of the applied field, the positive and negative ions in the crystal are displaced in opposite directions. The presence of defects or impurities in ionic crystals gives rise to additional absorption bands, producing peaks in the plot of dielectric loss ('$\tan \delta$') against frequency. If the loss mechanism due to the defect were characterized by a single energy, then the absorption line shape would be Lorenzian, varying with frequency ω according to the relation

$$\frac{\omega}{1+\omega^2\tau^2}$$

where $\tau = \tau_0 \exp(W/kT)$ is the relaxation time associated with the defect process. In practice, the peaks of $\tan \delta$ vs ω are found to be broader than the above relation indicates, showing that there are generally many relaxation processes occurring, with a range of relaxation times.

The first additional problem which arises in the interpretation of the dielectric properties of films of alkali halides is that the dielectric properties change with time after the films have been produced. The changes are for the most part restricted to frequencies below about 10 kHz, as shown in fig. 8.13, giving the variation of tan δ with time

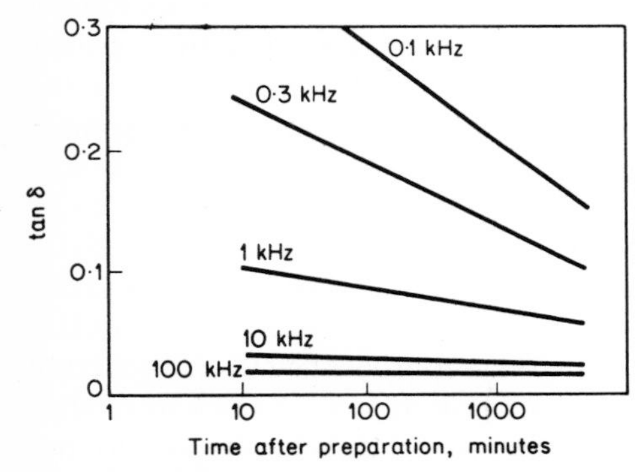

FIG. 8.13. Evolution of dielectric behaviour of sodium chloride films.

after formation of a film of NaCl, 260 nm thick (Weaver, 1966). This behaviour is not very surprising inasmuch as it is known that films formed by thermal evaporation have a very high density of defects when first formed, which then decay with time to an equilibrium value. In the alkali halides, F-centres are known to form readily and very large concentrations ($>10^{19}$ cm^{-3}) have been observed in freshly formed films. However, the simple notion of a parallel-sided, randomly-orientated polycrystalline film, with an initially large concentration of defects cannot account for other observed features in the dielectric behaviour. In particular, the frequency-dependence of the dielectric loss and the influence of temperature on the loss spectrum exhibit a far more complex behaviour than the simple 'defect loss' model gives.

Certain general systematic features emerge from studies of the magnitude of ageing effects and of losses as a function of the sizes of anions and cations involved. These may be summarized as follows:

(*i*) Ageing effects *increase* as *cation* size increases but *decrease* as *anion* size *increases*;

(*ii*) Losses *increase* as *anion* size *increases* but *decrease* as *cation* size *increases*.

The latter behaviour suggests that the losses may be associated with migration of cation vacancies. It is known that alkali halide films contain appreciable concentrations of such vacancies but a difficulty of the simple migration hypothesis is that, from the order of magnitude of the cation vacancies to be expected, the d.c. conductivity which would be expected is much higher than that actually observed.

Electron microscope studies of alkali halide films grown under the conditions used for the dielectric studies (thermal deposition in an oil-pumped system at 10^{-5}–10^{-6} torr) suggest a structure in which flat crystal platelets form and, as film thickness increases, overlap one another. This gives rise to the possibility that the migration of vacancies through the film may be impeded at intercrystallite boundaries. This is in fact consistent with the observed ageing behaviour because although the films are known to contain equal numbers of anion and cation vacancies (Schottky defects), the mobilities of the anion vacancies are known to be much lower than those of the cations.

Further supporting evidence for the role of intercrystalline boundaries is found from studies in which the crystallite size is varied. Experiments on films deposited on heated substrates (which produces larger crystallites than deposition on cold substrates) result in sharper peaks in the $\tan\delta$ vs ω curve, occurring at lower frequencies. Also, a similar behaviour is observed when films, initially measured *in vacuo*, are exposed to the atmosphere. In the latter case, it is likely that adsorption of moisture and recrystallization will lead to a larger effective crystallite size.

Loss measurements in the region of very low frequencies (down to $\sim$0·01 Hz) indicate the presence of rather broad dispersion maxima in the neighbourhood of 1 Hz and also of $\sim$0·01 Hz). Coupled with these results is the observation of a very large real part of the permittivity. It has been suggested that the blocking of the migration of vacancies, which so conveniently accounts for the ageing and (high-frequency) loss behaviour of the films may in fact not be total in the sense that, for sufficiently long times (i.e. for sufficiently low frequencies) migration through grain boundaries can occur. In the case of films of NaCl, the activation energy corresponding to the very low frequency peak is slightly higher than would be expected for a 'straight' migration process. It is plausible, however, that the intercrystallite boundary would present a somewhat higher potential barrier than that normally met by the migrating vacancy within the crystallites.

ANODIC OXIDE FILMS

The main interest in anodic films centres on those formed on aluminium and tantalum. Both materials have been successfully used for the manufacture of electrolytic capacitors. Structurally, anodic films formed on these metals are, at least in thicknesses up to hundreds of μm, amorphous. Dielectrically they are characterized, over the audio-frequency region, by almost constant values of $\tan\delta$. In a general way, one would certainly not expect an amorphous material to indicate sharp peaks in the $\tan\delta$ spectrum. The structural 'diffuseness' of such layers would indicate a spectrum of relaxation times. It is, however, mildly surprising that the loss spectrum is so featureless. Constancy of the value of $\tan\delta$ would result from a distribution $f(\tau)$ of relaxation times given by $f(\tau) = 1/\tau$. Such a distribution would result from a structure in which ions can make transitions between 'regular' sites, characterized by a fixed energy, and a distribution of sites with energies W such that the density of sites is proportional to $\exp(-W/kT)$. This empirical, but plausible, model was put forward by Garton (1946) and leads to a constant value of $\tan\delta$ for the oxides in question, for which the real part of the permittivity does not vary significantly with frequency.

The difficulty with amorphous films is that it is almost impossible to obtain structural information in sufficient detail even to make the crudest of guesses at the likely band structure, so that mainly indirect methods are needed in devising satisfactory models.

Very low frequency measurements on anodic oxide films do not reveal peaks in the $\tan\delta$ spectrum, such as are observed in alkali halide films. Typical results for Ta_2O_5 films (in the form of electrolytic capacitors) are shown in fig. 8.14.

DIELECTRIC FILMS BY OTHER PROCESSES

Although the electrolytic capacitor has proved an invaluable part of our electronic hardware for over a quarter of a century, the development of microelectronic technology has led to the requirement that capacitors or their insulating layers be formed *in situ* on a range of devices such as silicon monolithic circuits, thin-film transistors and thin-film memory devices. As a result, studies have

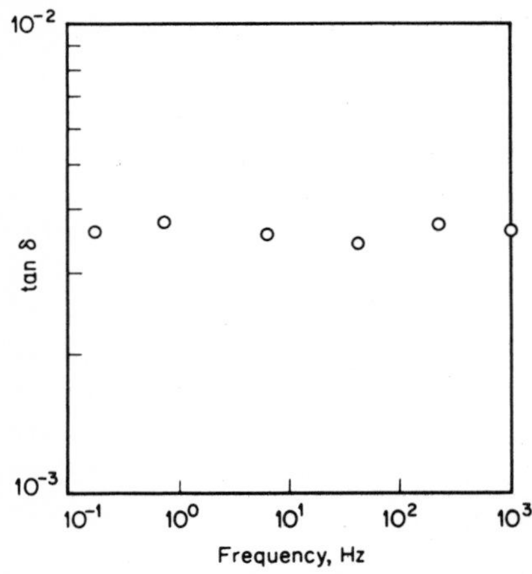

FIG. 8.14. Invariance of dielectric loss with frequency for Ta_2O_5 film.

been made of the dielectric properties of films formed by processes other than those of thermal evaporation or anodic oxidation. Two general classes of films have been examined, namely:

(*i*) Oxides, including mixed oxides;
(*ii*) Organic polymer films.

For studies of oxide films, much use has been made of the method of reactive sputtering which not only enables films of accurately controlled thickness to be formed but also, by adjustment of the composition of the sputtering gas, provides considerable control over the electrical properties of the films formed. Thus films of tin oxide made in this way cap be made as quite respectable conductors (sufficiently so for use as demisting coatings on windscreens) or, by the use of a high oxygen pressure during formation, as good insulators. For the production of mixed oxides, it proves easy to adjust the composition of the oxide by varying the relative areas of the two metals forming the cathode. Certain double oxides, such as lead titanate, are made in this way and have extremely high permittivities

($\sim$35). Moreover, the permittivity is found to be practically independent of temperature over a useful range around room temperature. The dielectric strength, at $\sim$400 kV.cm^{-1} is on the low side for some applications, for which films of SiO_2, or even of glass, would be used. In these cases, strengths of 3–5 MV.cm^{-1} are attainable, with capacitances of the order 0·03 μF.cm^{-2} for films 100 nm thick.

An alternative to sputtering, for films of SiO_2, is the method involving pyrolytic decomposition of a suitable compound of Si, either simply by the influence of heat or by reaction with another material. Thus tetraethoxysilane [$Si(OC_2H_5)_4$] decomposes at 750° C, producing SiO_2. If oxygen is used as a carrier gas, a lower temperature suffices (300° C) and clear, glassy films of the oxide form. This is but one of many silicon esters which behave in this way. For the production of films of the nitride of silicon (Si_3N_4), the pyrolytic reaction between silicon tetrafluoride and ammonia may be used, yielding films with very high resistivity (10^{14} ohm-cm) and a breakdown strength in excess of 10 MV.cm^{-2}. Films of this material have been used in the production of silicon field-effect transistors.

In the early days of electron diffraction and microscopy, in systems using oil as the pump fluid, insulating films were observed to build up on parts of the system in the neighbourhood of the electron beam. Nuisance though this was, the formation of films in this way – by electron bombardment of the oil vapour – has emerged as one of the possible techniques for insulating film production. The mechanism is that of polymerization, by the electron beam, of the fragments of the oil molecules. The method has been used to produce films of polymerized epoxy resins, yielding layers with permittivities of the order 5·3–6·2 with reasonably low loss factors (0·004–0·007). The dielectric strengths observed were in the neighbourhood of 1–2 MV.cm^{-1}. An interesting example of the need for nucleating sites for crystal growth arises in the case of polymerized epoxy films such as these. When normally prepared, they cannot be coated with metallic layers – the metal simply will not condense on the surface. After exposure to ultraviolet radiation, normal deposition of the metal occurs, suggesting that nucleating sites, formed by radiation damage of the polymer surface, are required for metal film growth.

An alternative to the use of electron beams for polymerization is that of using ultraviolet radiation, the process being known as photolysis. This process has been extensively used for the production of microcircuits, whereby an image of a desired circuit configuration repeated many times, is projected on to a layer of photo-resist on the silicón chip on which the circuit is to be formed. In one type of resist, the effect of the incident light is to render the exposed part of the resist insoluble, whereas the unexposed parts may be dissolved away by a suitable solution. Thus certain parts of the silicon surface are uncovered by the process and can subsequently be treated to form one or other microcircuit components.

Photo resist films of this kind are insulators, although their insulating characteristics are not impressive. They have, however, been used extensively in cryogenic thin-film memory circuits, suggesting that they possess good low-temperature behaviour.

Photolytic methods for the production of capacitors have been applied to methyl methacrylate, acrolein and divinylbenzene (Gregor, 1966). The polymer films so formed have permittivities in the region 3·2–3·5 and reasonably low loss factors provided films of sufficient thickness (>50 nm) are used. Breakdown field strengths in the neighbourhood of 3–5 MV.cm^{-1} are observed and the films are found to be stable over a wide temperature range (77 to 300° K). In some cases, however, loss peaks occur at low ($\sim$kHz) frequencies, possibly due to the presence of unpolymerized material in the films.

It seems possible that, despite some of the shortcomings of present dielectric films produced by processes such as those described above, these methods may well be developed to the point of becoming normal processes for thin-film microcircuitry.

References

CHAPTER 2

CABRERA N. and MOTT N. F., *Reports on Progress in Physics* **12**, 163 (1949).
HOLLAND L., 'Vacuum Deposition of Thin Films', Chapman & Hall Ltd. London (1956).

CHAPTER 3

SCHLIER R. E. and FARNSWORTH H. E., *Advances in Catalysis* **9**, 434 (1957).
THOMAS G., 'Transmission Electron Microscopy of Metals', Wiley, NY (1962).

CHAPTER 4

BALL D. J. and VENABLES J. A., *J. Vac. Sci. Tech.* **6**, 468 (1969).
CHAMBERS A. and PRUTTON M., *Thin Solid Films* **1**, 393 (1967).
HIRTH J. P. and POUND G. M., 'Single Crystal Films', Eds. Francombe and Sato, Pergamon Press, Oxford (1964).
LEWIS B. and CAMPBELL D. S., *J. Vac. Sci. Tech.* **4**, 209 (1967).
RHODIN T. N. and WALTON D., 'Single Crystal Films', Eds. Francombe and Sato, Pergamon Press, Oxford (1964).
WALTON D., *J. Chem. Phys.* **37**, 2182 (1962); *Phil. Mag.* **7**, 1671 (1962).

CHAPTER 5

BEAMS J. W., 'Structure and Properties of Thin Films', Wiley, NY (1959).
HOFFMAN R. W., 'The Use of Thin Films in Physical Investigations', Academic Press, NY (1966).

CHAPTER 6

ABELES F., *J. Phys. Rad.* **11**, 310 (1950).
ARCHER R. J., *J. Opt. Soc. Amer.* **52**, 970 (1962).
HEAVENS O. S., 'Physics of Thin Films', Vol. 2 (1964).
JACOBSSON R., *Opt. Act.* **10**, 309 (1963).

MALÉ D., *C.R. Acad. Sci. (Paris)* **230**, 1349 (1950).
SMITH D. O., *Opt. Act.* **12**, 13 (1965).
TEEGARDEN K. J., 'Optical Properties of Dielectric Films', Ed. N. N. Axelrod, Electrochem. Soc., NY (1968).

CHAPTER 7

CORCIOVEI A., *J. Phys. Chem. Solids*, 162 (1961).
GLASS S. J. and KLEIN M. J., *Phys. Rev.* **109**, 288 (1958).
NEEL L., *Compt. Rend.* **237**, 1468 (1953); *J. Phys. Rad.* **15**, 225 (1954).
NEUGEBAUER C. A., *Phys. Rev.* **116**, 1441 (1959).
SMITH D. O., *J. Appl. Phys.* **30**, 264S (1959).
TANIGUCHI S., *Sci. Rpts. Res. Insts., Tohoku Univ.* **A1**, 269 (1955).
VALENTA L., *Czech. J. Phys.* **7**, 127, 133 (1957).

CHAPTER 8

BARDEEN J., COOPER L. N. and SCHRIEFFER J. R., *Phys. Rev.* **108**, 1175 (1957).
FUCHS K., *Proc. Camb. Phil. Soc.* **34**, 100 (1938).
GARTON C. G., *Trans. Far. Soc.* **42A**, 56 (1946).
GINZBURG V. L. and LANDAU L. D., *J.E.T.P. (U.S.S.R.)* **20**, 1064 (1950).
GREGOR L. V., 'Physics of Thin Films', Vol.3., Eds. G. Hass and R. E. Thun, Academic Press, NY (1966).
HOLLAND L. and SIDDALL G., *Vacuum* **3**, 388 (1953).
HILL R. M., *Proc. Roy. Soc.* **A309**, 377, 397 (1969).
ITOH T., SHINICHI H. and KAMINAKA N., *J. Appl. Phys.* **40**, 2597 (1969).
LOCK J. M., *Proc. Roy. Soc.* **A208**, 391 (1951).
LONDON F. and LONDON H., *Proc. Roy. Soc.* **A149**, 71 (1935); *Physica*, **2**, 341 (1935).
PIPPARD A. B., *Proc. Roy. Soc.* **A203**, 210 (1950).
RICHARDS P. L. and TINKHAM M., *Phys. Rev.* **119**, 575 (1960).
SONDHEIMER E. H., *Adv. Phys.* **1**, 1 (1952).
WEAVER C., 'The Use of Thin Films in Physical Investigations', Ed. J. C. Anderson, Academic Press, NY (1966).
ZAVARITSKII N. V., *Dokl. Navk. SSR.* **78**, 665 (1950).
ZEMEL J. N., 'The Use of Thin Films in Physical Investigations', Ed. J. C. Anderson, Academic Press, NY (1966).

Index

BELOIT COLLEGE
530.41 H352t
Heavens, O. S. cn
Thin film physics by O.S. He
000
010101
0 2021 0058505 0